Progress in Nonlinear Differential Equations and Their Applications

Volume 33

I. Kuzin
S. Pohozaev

Entire Solutions of Semilinear Elliptic Equations

Birkhäuser
Basel · Boston · Berlin

I. Kuzin and
S. Pohozaev
Steklov Mathematical Institute
42 Vavilova
117966 Moscow
Russia

1991 Mathematics Subject Classification 35J60, 35K57, 35Q55, 35Q40

A CIP catalogue record for this book is available from the Library of Congress, Washington D.C., USA

Deutsche Bibliothek Cataloging-in-Publication Data
Kuzin, Il'ja A.:
Entire solutions of semilinear elliptic equations / I. Kuzin ;
S. Pohozaev. – Basel ; Boston ; Berlin : Birkhäuser, 1997
(Progress in nonlinear differential equations and their applications ; Vol. 33)
ISBN 3-7643-5323-6 (Basel ...)
ISBN 0-8176-5323-6 (Boston)

Printed on acid-free paper produced from chlorine-free pulp. TCF ∞
Printed in Germany
ISBN 3-7643-5323-6
ISBN 0-8176-5323-6

9 8 7 6 5 4 3 2 1

Contents

Introduction

This book is devoted to the study of elliptic semilinear equations of the type

$$-\Delta u + f(\mathbf{x}, u) = 0 \qquad \text{on } \mathbb{R}^N, \tag{0.1}$$

where $f : \mathbb{R}^N \times \mathbb{R} \to \mathbb{R}$ is a function of appropriate regularity. Equations of this type have many applications. For instance, many natural processes can be described by evolution equations

$$u_t - \Delta u + f(\mathbf{x}, u) = 0 \qquad \text{on } \mathbb{R}^+ \times \mathbb{R}^N, \tag{0.2}$$

where u is a real-valued function. Such equations are called reaction-diffusion equations and stationary states of (0.2) are described by (0.1).

Equations of the kind (0.1) arise also in nonlinear nonrelativistic quantum physics, where the nonlinear Schrödinger equation in a complex wave function u,

$$i\, u_t + \Delta u + f(\mathbf{x}, |u|)u = 0 \qquad \text{on } \mathbb{R} \times \mathbb{R}^N, \tag{0.3}$$

is used.

The standing wave type solutions of this equation,

$$u = e^{-iEt}\, v,$$

where v is a real-valued function, have an important physical interpretation. They correspond to stable states at the definite energy level E. Such states satisfy the equation, which is actually an equation of the type (0.1).

Then, the model equation describing scalar fields

$$u_{tt} - \Delta u + f(|u|)u = 0 \qquad \text{on } \mathbb{R} \times \mathbb{R}^N \tag{0.4}$$

is widely used in relativistic quantum physics. It is known as a nonlinear Klein-Gordon equation.

Of special interest among the solutions of (0.4) are so-called traveling waves, i.e., solutions of the type

$$u = v(\mathbf{x} - \mathbf{c}\, t),$$

where $\mathbf{c}$ is some vector in $\mathbb{R}^N$, $|\mathbf{c}| < 1$. Let v be real. It may be easily seen in Strauss [1] that, after a Lorentz transformation of coordinates, v satisfies an equation of form (0.1).

In quantum field theory the formal replacing of real time by imaginary time: $t \to i\,t$ is widely used. Without going into details about the physical meaning of such a substitution let us note that after this substitution equation (0.4) looks like (0.1) on $\mathbb{R}^{N+1}$. Its solutions with finite "energy" are called instantons and their existence or absence in a physical model is substantially important in the calculation of quantum processes.

Of course, the aforementioned problems do not exhaust the number of physical ones using equation (0.1) on $\mathbb{R}^N$. It is necessary to note that equations of the type (0.1) in bounded domains have been studied extensively (see, for instance, Gilbarg and Trudinger [1]; Ladyzhenskaya and Ural'ceva [1]). However, setting up a problem in an unbounded domain introduces new specific features. That is why such problems on $\mathbb{R}^N$ are of special mathematical interest.

The following is an example of such peculiarities. It is well known that if a smooth open domain $\Omega \subset \mathbb{R}^N$ is bounded, then for any $1 \le p < 2^*$, where $2^* = +\infty$ for $N \le 2$, and $2^* = 2N/(N-2)$ for $N > 2$, the compact embedding of corresponding functional spaces, $H_0^1(\Omega) \subset\subset L_p(\Omega)$, takes place. This fact is crucial for the investigation of nonlinear elliptic equations in such domains. However, if a domain is not bounded, compactness of embedding fails, which causes essential difficulties. These difficulties are intrinsic rather than cosmetic in character and this leads to completely different properties of nonlinear elliptic problems on $\mathbb{R}^N$.

Let us consider two examples

$$\begin{aligned} -\Delta u + |u|^{p-2}u &= 0 \qquad \text{in } \Omega, \\ u &= 0 \qquad \text{on } \partial\Omega, \end{aligned} \tag{I}$$

$$-\Delta u + |u|^{p-2}u = 0 \qquad \text{on } \mathbb{R}^N, \tag{II}$$

where Ω is a smooth open bounded domain in $\mathbb{R}^N$, $N \ge 3$. We will be interested in classical solutions, $u \in C^2(\Omega) \cap C(\bar{\Omega})$ or $u \in C^2_{\text{loc}}(\mathbb{R}^N)$. It is known that

I) If $2 < p < 2^*$, problem (I) possesses a positive solution and infinitely many solutions which change sign. On the other hand, for $p \ge 2^*$ and a wide class of domains (so-called star-shaped ones) this problem has no solution (Pohozaev [1]).

II) If $2 < p < 2^*$, any nonnegative solution of (II) is trivial, $u(\mathbf{x}) \equiv 0$ (Gidas and Spruck [1]). For $p \ge 2^*$, however, problem (II) possesses a continuum of decaying solutions, $u \to 0$ as $|\mathbf{x}| \to 0$, which can not be transferred to each other by translation.

All this generates a need for the development of new techniques and it is this need to which the present book is devoted.

The main goal of the book is the description of some existing methods for studying elliptic problems on $\mathbb{R}^N$. These methods are demonstrated where possible on simple examples. Of course, these examples do not exhaust the possibilities of the described methods. For instance, such methods could be also applied to quasilinear equations of a more general type

$$-\sum_{i=1}^{N} \nabla_i(|\nabla u|^{p-2}\nabla_i u) + f(\mathbf{x}, u) = 0 \qquad \text{on } \mathbb{R}^N,$$

etc.

Let us summarize briefly the book's contents.

Chapter 1 is devoted to the classical variational methods. §1 is a list of definitions and specific notation used in the book. The connection between classical solutions of (0.1) and so-called generalized ones is also discussed. Some more details concerning this matter can be found in Appendix A.

In §2 the classical variational method is formulated in an abstract form. This method is then demonstrated on some elliptic problems. Problems described here show the specific role of the nonlinearity exponent 2^*.

In §3 the possibility to approximate a solution on $\mathbb{R}^N$ by solutions to appropriate problems defined in bounded domains is demonstrated. In §4 the variational method is applied to problems in balls; resulting solutions approximate the solution on $\mathbb{R}^N$ according to §3.

In §5 we consider the simplest application of the classical monotonicity method to an elliptic problem on $\mathbb{R}^N$.

More complicated so-called noncoercive problems with a parameter (eigenvalue) are considered in Chapter 2 with the use of the conditional extremum problems.

§6 contains the proofs of abstract theorems applied in §§7–10 to some elliptic problems. For this purpose in §9 the results of Berestycki and Lions [1] and in §10 some ideas of Rother [1] and Stuart [1] are used.

§11 is the first example of the compactification method proposed by Lions [2,3] and known as the concentration compactness method. In another context this method was introduced by Ding and Ni [1] and may be called the comparison method. The main idea of this compactification originates from a comparison of two appropriate variational problems. §12 discusses homogeneous problems.

Chapter 3 is devoted to more complicated methods.

In §13 the first simplest application of the mountain pass theorem by Ambrosetti and Rabinowitz [1] to noncoercive problems on $\mathbb{R}^N$ is given. The rudiments of the mountain pass method can be found in Appendix E.

§§14–16 show how the mountain pass method together with the concentration compactness (comparison) method can be applied to noncoercive elliptic problems on $\mathbb{R}^N$. We follow here, in particular, Lions [2,3], Ding and Ni [1], Benchi Cerami [1], Zhu Xi-Ping and Cao Daomin [1].

§17 provides an example of variational problems in invariant spaces different from spherically symmetric.

In §§18 the ideas of §§3 and 4 are developed using more complicated concepts.

In Chapter 4 the ODE method (which, of course, can be applied only to the radial solutions) is introduced. Basic theorems of this method are recalled in §19.

In §20 the rather detailed classification of the solution behavior on $\mathbb{R}^N$ for autonomous elliptic equations is given. Here the results of Keller [1] are presented. The theorems of §21 exhaust the Dirichlet problem for the one-dimensional case. Here we use the work of Berestycki and Lions [1].

§22 demonstrates the simplest ideas of phase plane and dynamic systems applied to the nonlinear autonomous problem on $\mathbb{R}^N$ according to Chandrasekhar

[1] and Joseph and Lundgren [1]. §23 is a simple exercise in substitution that proves to be useful sometimes. §24 presents the shooting method, which is based on the continuous dependence of solutions on the initial data. This is a simplified version of the paper by Berestycki, Lions, and Peletier [1].

Chapter 5 is devoted to other modern methods.

§25 yields the approach of upper and lower solutions together with simple examples. Here we use the paper by Fukagai [1].

§26 is no more than an illustration of the Leray-Schauder method applied to the specific problems arising on $\mathbb{R}^N$.

§27 contains some special integral identities and their applications to elliptic problems on $\mathbb{R}^N$.

§28 is devoted to the presentation of the fibration method in an abstract form. It makes it possible, in particular, to prove the existence of infinitely many solutions to corresponding elliptic problems.

§29 provides examples of nonexistence theorems.

Some methods have been omitted in this book. To begin with, we did not consider anticoercive problems where the critical nonlinearity exponent arises. In this case difficulties can already be seen in bounded domains, and our goal was to focus on the problems specific to $\mathbb{R}^N$. We also omitted more delicate methods of dynamic systems (Jones and Kupper [1], Jones, Kupper, and Plaktier [1], Budd and Norburg [1]), uniqueness problems for positive solutions (Peletier and Serrin [1], McLeod and Serrin [1], Kwong [1]), singularities removing (Brezis and Veron [1], Veron [1]), fine methods of the maximum principle (Gidas, Ni, and Nirenberg [1], Gidas and Nirenberg [1]), problems of multiple solvability (Berestycki and Lions [2], Xi-Ping Zhu [1]), etc. We did so because of, first, the limited volume of the book and, second, our wish to concentrate on the methods that are sufficiently clear and complete at present.

The literature on entire solutions is huge, and we did not even try to review it. We were forced to restrict ourselves to those references whose results or approaches were directly used in the book or which were very close to the matter considered.

We apologize to all the authors whose works and merit in this field occured beyond our consideration.

We are very grateful to the editor Dr. T. Hintermann whose remarks improved considerably the text of this book.

§0. Notation

Throughout the book, the following conventions are used.

1. By bold **C** we mean different positive constants, which in every particular case are absolute, i.e., do not depend on n, u, etc. As a rule, we will not stress this fact but it will follow from the context.

2. Sequences $\{u_n\}$ and their subsequences will be denoted equally $\{u_n\}$ in order to avoid cumbrous subindexes.

Chapter 1
Classical Variational Method

In this chapter and throughout the book, we shall consider boundary problems of the form

$$
\begin{aligned}
-\Delta u + f(\mathbf{x}, u) = 0 \qquad & \text{on } \mathbb{R}^N, \\
u \to 0 \qquad \text{as } |\mathbf{x}| \to \infty. &
\end{aligned}
\tag{1.1}
$$

Functions $f : \mathbb{R}^N \times \mathbb{R} \to \mathbb{R}$ are supposed to satisfy a so-called Carathéodory condition, i.e., to be continuous with respect to t for almost all $\mathbf{x} \in \mathbb{R}^N$ and measurable in $\mathbf{x}$ for all $t \in \mathbb{R}$.

§1. Preliminaries

It is convenient in the variational approach, to consider not only classical solutions $u \in C^2_{\text{loc}}(\mathbb{R}^N)$, that is, solutions with continuous second derivatives, but a wider class of so-called generalized solutions belonging to wider functional spaces. In the present section, we define these spaces and describe in what sense generalized solutions to problem (1.1) are understood.

Actually, all results presented here are classical. It is only worth mentioning that our definitions are often more narrow because they are necessary only in applications to indicated problems. You can find more complete information concerning functional spaces and functional analysis, for example, in the following manuals: Berger [2], Dunford and Schwartz [1], Yosida [1].

1.1 Main Functional Spaces

Let Ω be a bounded or unbounded domain in $\mathbb{R}^N$. In this book, Ω is actually $\mathbb{R}^N$ or B_R, where we denote by B_R a ball of the radius R centered at the origin of $\mathbb{R}^N$.

Let $\mathcal{D}(\Omega)$ be the set of infinitely differentiable functions whose support is compact in Ω. Define functionals

$$
\|u\|_p(\Omega) = \left(\int_\Omega |u(\mathbf{x})|^p \, d\mathbf{x} \right)^{1/p}, \qquad 1 \le p < +\infty,
$$

on $\mathcal{D}(\Omega)$. Also define

$$
\|u\|_\infty(\Omega) = \sup_{\mathbf{x} \in \Omega} |u(\mathbf{x})|.
$$

If $\Omega = \mathbb{R}^N$, we omit it in the notation of norms.

Recall some definitions. Functions from $\mathcal{D}(\Omega)$ together with their closure with respect to the norm

$$\|u\|_p(\Omega), \qquad 1 \leq p < \infty,$$

form the Banach space of Lebesgue which is denoted as $L_p(\Omega)$. Analogously, closure of $\mathcal{D}(\Omega)$ with respect to the norm

$$\|\nabla u\|_2(\Omega) + \|u\|_2(\Omega)$$

forms the Banach space of Sobolev denoted as $H_0^1(\Omega)$.

For the sake of convenience, we introduce also the following notation. Denote the closure of $\mathcal{D}(\mathbb{R}^N)$ in the norm

$$\|\nabla u\|_2 + \|u\|_p + \|u\|_q, \qquad 1 \leq p \leq q < \infty,$$

$\mathcal{E}_{p,q}$; also let $\mathcal{E}_{p,p} = \mathcal{E}_p$. In particular, $\mathcal{E}_2 = H_0^1(\mathbb{R}^N)$.

Define

$$2^* = \begin{cases} +\infty & \text{if } N \leq 2, \\ \frac{2N}{N-2} & \text{if } N > 2. \end{cases}$$

The number 2^* is called Sobolev's critical exponent.

We say that a Banach space X is continuously embedded into a Banach space Y (which is denoted $X \circlearrowleft Y$) if $u \in X$ implies $u \in Y$ and there exists a constant $\mathbf{C}$ not depending on u such that for all $u \in X$,

$$\|u\|_Y \leq \mathbf{C}\|u\|_X,$$

where $\|.\|$ are the corresponding norms.

The theory of functional spaces provides the following embedding theorems.

SOBOLEV EMBEDDING THEOREM. *Let $N > 2$. Then for all exponents $1 \leq p \leq q \leq 2^*$,*

$$\mathcal{E}_p \circlearrowleft L_q(\mathbb{R}^N).$$

SOBOLEV EMBEDDING THEOREM. *Let $N > 2$. Then for all exponents $2^* \leq p \leq q < +\infty$,*

$$\mathcal{E}_q \circlearrowleft L_p(\mathbb{R}^N).$$

The Sobolev spaces are discussed in detail in Adams [1].

1.2 Weak and Generalized Solutions and Their Connection with Classical Ones

Consider equation (1.1) without a boundary condition

$$-\Delta u + f(\mathbf{x}, u) = 0 \qquad \text{on } \mathbb{R}^N. \tag{1.2}$$

Recall that throughout the book we assume that f is a Carathéodory function.

DEFINITION. A locally measurable function $u : \mathbb{R} \to \mathbb{R}^N$ is a weak solution of (1.2) if functions ∇u and $f(\,.\,, u(.))$ are locally integrable and for all $\varphi \in \mathcal{D}(\mathbb{R}^N)$ the identity

$$\int_{\mathbb{R}^N} (\nabla u, \nabla \varphi)\, d\mathbf{x} + \int_{\mathbb{R}^N} f(\mathbf{x}, u)\varphi\, d\mathbf{x} = 0$$

holds.

DEFINITION. A twice continuously differentiable function, $u \in C^2_{\text{loc}}(\mathbb{R}^N)$, is called a classical solution to equation (1.2) if for this u (1.2) holds identically at each point $\mathbf{x} \in \mathbb{R}^N$.

The connection between classical and weak solutions is well known.

THEOREM 1.1. *Any classical solution $u \in C^2_{\text{loc}}(\mathbb{R}^N)$ to equation (1.2) is its weak solution.*

Proof. It is sufficient to multiply (1.2) by $\varphi \in \mathcal{D}(\mathbb{R}^N)$ and integrate by parts. □

Obtaining the opposite direction is more difficult and is the subject of a separate investigation (e.g., Gilbarg and Trudinger [1], Ladyzhenskaya and Ural'ceva [1]). We adduce here some of the results without proof. Because the regularity of weak solutions depends on the characteristics of f, we shall impose on f some additional conditions, which we formulate below.

CONDITION D1. For any ball $B_R \subset \mathbb{R}^N$ there exists a continuous function $\psi : \mathbb{R} \to \mathbb{R}$ with

$$\lim_{|t|\to\infty} \frac{\psi(t)}{|t|^p} = 0 \quad \text{for some } 1 < p < +\infty \qquad \text{if } N \le 2,$$

$$\lim_{|t|\to\infty} \frac{\psi(t)}{|t|^{2^*}} = 0 \qquad \text{if } N > 2$$

such that

$$t\, f(\mathbf{x}, t) \ge \psi(t)$$

holds for almost all $\mathbf{x} \in B_R$ and for all $t \in \mathbb{R}$.

CONDITION D2. The function f is locally bounded, that is, for any positive constants R and M there exists a constant A_M depending on M such that for almost all $\mathbf{x} \in B_R$ and for all $|t| \leq M$,

$$|f(\mathbf{x}, t)| \leq A_M.$$

Recall that a function $g : \mathbb{R}^N \to \mathbb{R}$ is called locally Hölder on $\mathbb{R}^N$ if it is continuous on $\mathbb{R}^N$ and for any ball B_R there exists a constant $\alpha \in (0, 1)$ such that

$$\sup_{\substack{\mathbf{x} \in B_R \\ \mathbf{y} \in B_R \\ \mathbf{x} \neq \mathbf{y}}} \frac{|g(\mathbf{x}) - g(\mathbf{y})|}{|\mathbf{x} - \mathbf{y}|^\alpha} < +\infty.$$

We continue the formulation of conditions imposed on f.

CONDITION D3. The function f is such that for any function $w \in C^1_{\mathrm{loc}}(\mathbb{R}^N)$ the function $f(\,.\,, w(.)) : \mathbb{R}^N \to \mathbb{R}$ is locally Hölder on $\mathbb{R}^N$.

Imposing on f corresponding conditions, we get the theorems consecutively improving the regularity of a weak solution.

THEOREM 1.2. *Let a Carathéodory function f satisfy D1. Let u be a weak solution of (1.2) such that for any B_R there exists a constant C such that*

$$\|\nabla u\|_2(B_R) + \|u\|_1(B_R) \leq C. \tag{1.3}$$

Then for any B_R,

$$\|\nabla u\|_2(B_R) + \|u\|_\infty(B_R) \leq C' \tag{1.4}$$

holds, where C' does not depend on u.

Remark. If constants p, A_M, C, and the function ψ do not depend on R, then C' also does not depend on R.

THEOREM 1.3. *Let f satisfy D1 and D2. Let u be such a weak solution of (1.2) that for any B_R, (1.4) is valid. Then $u \in C^1_{\mathrm{loc}}(\mathbb{R}^N)$ and for any B_R,*

$$\sup_{\mathbf{x} \in B_R} |\nabla u| \leq C''$$

holds, where C'' does not depend on u.

Remark. If constants A_M in Condition D2 and C' in (1.4) do not depend on R, then C'' also does not depend on R.

THEOREM 1.4. *Let f satisfy D1, D2, and D3. Then any weak solution u of (1.2) satisfying (1.3) belongs to $C^2_{\mathrm{loc}}(\mathbb{R}^N)$.*

Until now the results presented have been of "an intrinsic character" and say nothing about the behavior of u as $|\mathbf{x}| \to +\infty$. In order to make sense of the boundary condition we introduce the notion of a generalized solution.

DEFINITION. A weak solution u to (1.2) is called a generalized solution to (1.1) if $u \in \mathcal{E}_p$ for some $1 \le p < +\infty$.

Later on we shall say a solution $u \in \mathcal{E}_p$, always meaning a generalized solution.

THEOREM 1.5. *Let $u \in \mathcal{E}_p$ be a uniformly bounded weak solution of (1.2) and a Carathéodory f be such that the function $f(\,.\,, u(\,.\,))$ is uniformly bounded on $\mathbb{R}^N$. Then, uniformly,*

$$u \to 0 \qquad \text{as } |\mathbf{x}| \to \infty. \tag{1.5}$$

Remark. If (1.5) holds, we shall say that the boundary condition of (1.1) is satisfied in the classical sense. Hence any generalized solution to (1.2), with f satisfying assumptions of Theorems 1.2–1.5, is a classical solution to boundary problem (1.1).

Proof of Theorem 1.5. Due to the conditions of the theorem, the assumptions of Theorem 1.3 hold, with constants not depending on R. Then

$$\sup_{\mathbf{x}\in\mathbb{R}^N} |\nabla u| \le C'' \tag{1.6}$$

for some constant C''. Let, contrarily, (1.5) be not true. Then there exist a constant $\varepsilon > 0$ and a sequence of points $\{\mathbf{x}_i\} \subset \mathbb{R}^N$, with $|\mathbf{x}_i| \to \infty$ as $i \to \infty$, such that

$$|u(\mathbf{x}_i)| \ge \varepsilon \tag{1.7}$$

for all i. Choose a subsequence from $\{\mathbf{x}_i\}$ such that

$$\inf_{i\neq j} |\mathbf{x}_i - \mathbf{x}_j| \ge \varepsilon/C''. \tag{1.8}$$

In accordance with (1.6) and (1.7)

$$|u(\mathbf{x})| \ge \varepsilon/2 \qquad \text{for all } \mathbf{x} \text{ with } |\mathbf{x} - \mathbf{x}_i| \le \frac{\varepsilon}{2C''}. \tag{1.9}$$

From (1.8) and (1.9)

$$\int_{\mathbb{R}^N} |u|^p\, d\mathbf{x} \ge \sum_{i=1}^{\infty} \int_{|\mathbf{x}-\mathbf{x}_i|\le \frac{\varepsilon}{2C''}} |u|^p\, d\mathbf{x} \ge \sum_{i=1}^{\infty} (\varepsilon/2)^p \operatorname{mes} B_{\varepsilon/2C''} = +\infty,$$

which contradicts $u \in L_p(\mathbb{R}^N)$ and thus proves the theorem. □

Remark. We do not investigate the specific behavior of u at infinity. We stress again than when we write $u \to 0$ as $|\mathbf{x}| \to \infty$, we mean this in a generalized sense. Theorem 1.5 justifies such an approach for many regular cases.

More definite information on the behavior of u at infinity may be found, for example, in Sobolev [1], Chapter 4, Uspenski [1], and Oleinik [1].

1.3 THE FUNCTIONAL OF THE PROBLEM

In order to define a functional corresponding to the problem under consideration, it is necessary to impose some conditions on f.

DEFINITION. Let $1 < p \leq q < +\infty$. We say that a Carathéodory function $f : \Omega \times \mathbb{R} \to \mathbb{R}$ belongs to the class $K_{p,q}(\Omega)$ if there exist a positive constant $\mathbf{C}$ and functions $f_1 \in L_{p/(p-1)}(\Omega)$, $f_2 \in L_{q/(q-1)}(\Omega)$ such that

$$|f(\mathbf{x}, t)| \leq \mathbf{C}(|t|^{p-1} + |t|^{q-1}) + f_1(\mathbf{x}) + f_2(\mathbf{x})$$

holds for almost all $\mathbf{x} \in \Omega$ and for all $t \in \mathbb{R}$. Denote $K_p(\Omega) = K_{p,p}(\Omega)$. If $\Omega = \mathbb{R}^N$, we omit it in the notation.

Let

$$F(\mathbf{x}, t) = \int_0^t f(\mathbf{x}, s)\, ds.$$

If $f \in K_{p,q}$, the functional

$$V(u) = \int_{\mathbb{R}^N} F(\mathbf{x}, u(\mathbf{x}))\, d\mathbf{x}$$

is defined on $L_p(\mathbb{R}^N) \cap L_q(\mathbb{R}^N)$. The functional

$$T(u) = \|\nabla u\|_2^2$$

is defined on $\mathcal{E}_p$ for all $1 < p < +\infty$. Hence, in this case, we can consider the functional

$$\Phi = \frac{1}{2}T + V$$

on the space $\mathcal{E}_{p,q}$.

The functional Φ is called the functional of problem (1.1).

It is easy to see (details in Appendix A) that $T \in C^1(\mathcal{E}_{p,q})$, and the value of the derivative $T'(u) \in \mathcal{E}^*_{p,q}$ on some element $v \in \mathcal{E}_{p,q}$ is expressed by

$$\langle T'(u), v\rangle = 2 \int_{\mathbb{R}^N} (\nabla u, \nabla v)\, d\mathbf{x}.$$

Analogously, because $f \in K_{p,q}$, $V \in C^1(\mathcal{E}_{p,q})$, and the value of the derivative $V'(u) \in \mathcal{E}^*_{p,q}$ on some element $v \in \mathcal{E}_{p,q}$ is expressed by

$$\langle V'(u), v\rangle = \int_{\mathbb{R}^N} f(\mathbf{x}, u)v\, d\mathbf{x}.$$

Obviously, $\Phi \in C^1(\mathcal{E}_{p,q})$ as well.

1.4 Critical Points. Basis of Variational Methods

The basis of the variational method applied to problem (1.1) is an abstract theorem that will be presented later. Let X be a Banach space and $J \in C^1(X)$ be a functional. Let X^* be dual to X (Appendix A).

DEFINITION. An element $u \in X$ is called a critical point of $J : X \to \mathbb{R}$ if

$$J'(u) = 0 \qquad \text{in } X^*,$$

i.e.,

$$\langle J'(u), v \rangle = 0$$

for all $v \in X$.

The next theorem follows immediately from this definition.

THEOREM 1.6. *Let $f \in K_{p,q}$, and $u \in \mathcal{E}_{p,q}$ be a critical point of the problem functional of (1.1). Then u is a generalized solution of this problem.*

There exists a result in the opposite direction.

THEOREM 1.7. *Let $u \in \mathcal{E}_{p,q}$ be a generalized solution of problem (1.1) with $f \in K_{p,q}$. Then u is a critical point of the problem functional $\Phi : \mathcal{E}_{p,q} \to \mathbb{R}$.*

Remark. Not any generalized or classical solution u of (1.1) is a critical point of the problem functional Φ considered on the space $\mathcal{E}_{p,q}$ naturally defined by $f \in K_{p,q}$.

As an example consider the following problem

$$-\Delta u + \frac{(8-2N)|\mathbf{x}|^2 - 2N}{(|\mathbf{x}|^2+1)^2} u = 0 \qquad \text{in } \mathbb{R}^N.$$

This problem has a classical solution $u_0 \in C^2(\mathbb{R}^N)$ for all N,

$$u_0(\mathbf{x}) = (|\mathbf{x}|^2 + 1)^{-1},$$

which is generalized for $N < 6$ in the sense of the definition of Section 1.2. We see that $f \in K_2$ and the problem functional Φ is defined on $H_0^1(\mathbb{R}^N) = \mathcal{E}_2$. On the other hand, $u_0 \notin H_0^1(\mathbb{R}^N)$ if $N > 3$.

In order to study this problem by variational methods we need more general so-called weighted spaces, which will be considered later. In this chapter we restrict ourselves only to aforementioned definitions.

§2. The Classical Method: Absolute Minimum

In accordance with what was said in the previous section we can solve problem (1.1) by finding critical points of the problem functional. In this section the classical and most simple variational method based on the theorem on a minimum of a functional is presented. When applied to differential equation (1.1), this method works well provided the functional possesses, in particular, a characteristic called coerciveness. We shall give a definition and present the corresponding abstract theorem.

We shall then demonstrate applications of the method using concrete examples. These examples may seem to be somewhat tedious but they are purposely detailed to demonstrate the role of the critical Sobolev exponent 2^*.

The classical variational method is contained in, among others, the following books: Berger [2], Krasnoselskii [1], Vainberg [1].

2.1 Foundations of the Method

The classical variational method is based on the following theorem on a critical point.

THEOREM 2.1. *Let X be a Banach space and a functional $J : X \to \mathbb{R}$ of the class C^1 be given. Let $u \in X$ be a point of an absolute minimum of J. Then*

$$J'(u) = 0 \qquad \textit{in } X^*,$$

where X^* is a dual space to X.

Following this theorem and Theorem 1.6 we may associate the problem functional $\Phi : \mathcal{E}_{p,q} \to \mathbb{R}$ with the original problem

$$-\Delta u + f(\mathbf{x}, u) = 0 \qquad \text{in } \mathbb{R}^N,$$

provided f belongs to $K_{p,q}$, and study the existence of a minimum of Φ.

The proof of the existence of a minimum is based on a general theorem, which we formulate here.

2.2 Basic Notions

Let X be a Banach space.

DEFINITION. A functional $J : X \to \mathbb{R}$ is called bounded on X from below if there exists a constant M such that

$$J(u) \geq M$$

for all $u \in X$.

DEFINITION. A functional $J : X \to \mathbb{R}$ is called weakly lower semicontinuous on X if one can choose from any weakly convergent sequence $\{u_n\} \subset X$,

$$u_n \rightharpoonup u \qquad \text{in } X,$$

a subsequence $\{u_n\}$ such that

$$\lim_{n\to\infty} J(u_n) \geq J(u).$$

DEFINITION. A functional $J : X \to \mathbb{R}$ is called coercive on X if for any constant M there exists a constant $C(M)$ such that for all $u \in X$ the inequality

$$|J(u)| \leq M$$

implies

$$\|u\|_X \leq C(M).$$

2.3 EXAMPLE: FUNCTIONAL T

As a useful example, consider the functional T defined in Section 1.3.

LEMMA 2.2. *The functional T is bounded from below on $\mathcal{E}_{p,q}$ for all $1 < p \leq q < +\infty$.*

Proof. This is obvious from $T(u) \geq 0$ for all $u \in \mathcal{E}_{p,q}$. □

LEMMA 2.3. *The functional T is coercive on $\mathcal{E}_{2^*}$ provided $N > 2$.*

Proof. Let $M > 0$ and $|T(u)| \leq M$. Then, by the Sobolev embedding theorem with $p = q = 2^*$,

$$\|\nabla u\|_2 + \|u\|_{2^*} \leq \mathbf{C}\,\|\nabla u\|_2 \leq \mathbf{C}M^{1/2} = \mathbf{C}(M)$$

for all $u \in \mathcal{E}_{2^*}$. □

LEMMA 2.4. *The functional T is weakly lower semicontinuous on $\mathcal{E}_{p,q}$ for all $1 < p \leq q < +\infty$.*

Proof. Let $\{u_n\}$ be a sequence from $\mathcal{E}_{p,q}$ satisfying

$$u_n \rightharpoonup u \qquad \text{in } \mathcal{E}_{p,q}. \tag{2.1}$$

We claim that

$$\underline{\lim}_{n\to\infty} T(u_n) \geq T(u).$$

Note that $0 \leq (a-b)^2$ for any real a and b, then

$$a^2 \leq b^2 + 2a(a-b).$$

Hence,

$$\|\nabla u\|_2^2 \leq \|\nabla u_n\|_2^2 + 2 \int_{\mathbb{R}^N} (\nabla u, (\nabla u - \nabla u_n))\, d\mathbf{x} \tag{2.2}$$

holds for any u, u_n from $\mathcal{E}_{p,q}$. From (2.1) we can take the limit in (2.2) and get

$$\begin{aligned} T(u) = \|\nabla u\|_2^2 &\leq \varliminf_{n\to\infty} \left(\|\nabla u_n\|_2^2 + 2 \int_{\mathbb{R}^N} (\nabla u, (\nabla u - \nabla u_n))\, d\mathbf{x} \right) \\ &= \varliminf_{n\to\infty} \|\nabla u_n\|_2^2 = \varliminf_{n\to\infty} T(u_n). \end{aligned}$$

□

2.4 THEOREM ON A MINIMUM OF A FUNCTIONAL

EXAMPLE. Consider the function $f : \mathbb{R}^+ \to \mathbb{R}^+, f(t) = 1/t$. It is easy to see that

$$\inf_{t\in\mathbb{R}^+} f(t) = 0,$$

but this function does not have a minimum, i.e., a point $t_0 \in \mathbb{R}^+$ with $f(t_0) = 0$. This means that the question of whether a functional attains its minimum is not trivial.

Recall that a Banach space X is called reflexive if $X^{**} = X$. For any Banach space X, $u_n \rightharpoonup u$ in X implies the boundedness of $\{u_n\}$ in X; if X is reflexive, then from any sequence bounded in X, one can choose a weakly convergent subsequence.

It is known that $\mathcal{E}_{p,q}$, with $1 < p, q < +\infty$, are reflexive.

The following is a theorem on a minimum of a functional.

THEOREM 2.5. *Let X be a reflexive Banach space. Let a functional $J : X \to \mathbb{R}$ be*

(i) bounded from below on X,
(ii) weakly lower semicontinuous on X,
(iii) coercive on X.

Then it has a minimum.

Proof. In accordance with (i) and the definition of an infimum there exists a minimizing sequence $\{u_n\} \subset X$, i.e., such a sequence that

$$\lim_{n\to\infty} J(u_n) = I, \tag{2.3}$$

where

$$I = \inf_{u \in X} J(u). \tag{2.4}$$

From (2.3) there exists a constant $\mathbf{C}$ such that

$$|J(u_n)| \leq \mathbf{C} \tag{2.5}$$

holds for all u_n. From the coerciveness of J and (2.5),

$$\|u_n\|_X \leq \mathbf{C}.$$

Because X is reflexive, we can choose a subsequence

$$u_n \rightharpoonup u \quad \text{in } X \qquad \text{as } n \to \infty.$$

Owing to the weak lower semicontinuousness of J, we have up to a subsequence

$$I = \lim_{n \to \infty} J(u_n) \geq J(u). \tag{2.6}$$

Because, according to definition (2.4),

$$J(u) \geq I \tag{2.7}$$

for all $u \in X$, we have, taking into consideration (2.6) and (2.7),

$$J(u) = I,$$

that is, the minimum I is achieved at the point $u \in X$. □

Remark. For the function f from the Example, this theorem does not work because f is not coercive (verify!).

The following sections are devoted to applications of this theorem.

2.5 THE FIRST SAMPLE

Now we can start the study of boundary problem (1.1), associating a variational one with it. In order to make use of Theorem 2.5 we need to know the properties of the problem functional. These properties depend substantially on the behavior of f both at infinity and at zero. In accordance with this we shall consider different conditions. Let us begin with a simple example:

$$-\Delta u + g(\mathbf{x}, u) = h(\mathbf{x}) \qquad \text{on } \mathbb{R}^N, \tag{P$_1$}$$

where

$$g \in K_p, \tag{g$_p$}$$

$$h \in L_{p/(p-1)}(\mathbb{R}^N) \tag{h$_p$}$$

for some $1 < p < +\infty$. It is easy to see that

$$f \in K_p, \quad \text{where} \quad f(\mathbf{x},t) = g(\mathbf{x},t) - h(\mathbf{x}).$$

Denote

$$G(\mathbf{x},t) = \int_0^t g(\mathbf{x},s)\,ds.$$

Suppose that there exist a constant $\mathbf{C} > 0$ and a function $\varphi \in L_1(\mathbb{R}^N)$ such that

$$G(\mathbf{x},t) \geq \mathbf{C}|t|^p + \varphi(\mathbf{x}) \tag{G$_p$}$$

for almost all $\mathbf{x} \in \mathbb{R}^N$ and for all $t \in \mathbb{R}$.

THEOREM 2.6. *Let conditions* (g_p), (G_p), (h_p) *be satisfied. Then problem* (P_1) *has a generalized solution* $u \in \mathcal{E}_p$.

Remark. Recall that in accordance with the theorems of Section 1.2, if the functions g and h from Theorem 2.6 are regular enough, the generalized solution u is classical, and the boundary condition is also satisfied in the classical sense.

EXAMPLE. The problem

$$-\Delta u + \lambda |u|^{p-2} u = h(\mathbf{x}) \quad \text{on} \quad \mathbb{R}^N,$$

has a generalized solution $u \in \mathcal{E}_p$ for any N, $1 < p < +\infty$, $\lambda > 0$, and $h \in L_{p/(p-1)}(\mathbb{R}^N)$.

Proof of Theorem 2.6. Consider the variational problem

$$\{u \in \mathcal{E}_p, \min \Phi(u)\} \tag{V$_1$}$$

where the problem functional $\Phi : \mathcal{E}_p \to \mathbb{R}$ has the form

$$\Phi(u) = \frac{1}{2}\|\nabla u\|_2^2 + V(u),$$

$$V(u) = \int_{\mathbb{R}^N} G(\mathbf{x},u)\,d\mathbf{x} - \int_{\mathbb{R}^N} h\,u\,d\mathbf{x}.$$

A solution of (V_1), if it exists, is a generalized solution of boundary problem (P_1) from Theorem 2.1 and Theorem 1.6. To prove solvability of variational problem (V_1) we have, in accordance with Theorem 2.5, to show that the functional Φ is

(i) bounded from below,
(ii) weakly lower semicontinuous,
(iii) coercive

on $\mathcal{E}_p$. If we prove these properties, the proof of Theorem 2.6 will be completed. □

2.6 KONDRASHOV THEOREM. LEBESGUE-FATOU THEOREM. HÖLDER INEQUALITY

Before starting the proof recall some well-known results from functional analysis, which we shall permanently use here and later on.

KONDRASHOV THEOREM. *Let $\{u_n\}$ be a sequence from $\mathcal{E}_p$ weakly convergent in $\mathcal{E}_p$ to some u. Then*

$$u_n \to u \qquad \text{in } L_q(B_R)$$

for all $1 \leq q < 2^*$ and all finite R.

LEBESGUE-FATOU THEOREM. *Let a sequence $\{u_n\} \subset L_1(\Omega)$ be uniformly bounded from below by a function $\varphi \in L_1(\Omega)$. Let*

$$\int_\Omega u_n \, d\mathbf{x} \leq \mathbf{C}.$$

Let, to this end,

$$u_n \to u \qquad \text{almost everywhere in } \Omega \qquad \text{as } n \to \infty.$$

Then u belongs to $L_1(\Omega)$, and

$$\int_\Omega u \, d\mathbf{x} \leq \varliminf_{n\to\infty} \int_\Omega u_n \, d\mathbf{x}\,.$$

HÖLDER INEQUALITY. *Let $u \in L_p(\Omega)$ and $v \in L_q(\Omega)$, with*

$$\frac{1}{p} + \frac{1}{q} = 1.$$

Then the function $w = uv$ belongs to $L_1(\Omega)$, and

$$\|w\|_1(\Omega) \leq \|u\|_p(\Omega)\,\|v\|_q(\Omega).$$

2.7 FIRST SAMPLE: THE PROOF

According to what has been stated above, we only need to check properties (i), (ii), (iii). Let us begin with property (iii).

Coerciveness of Φ. Choose M in such a way that the set

$$E \equiv \{u \in \mathcal{E}_p, |\Phi(u)| \leq M\}$$

is nonempty. From the Hölder inequality

$$\left|\int_{\mathbb{R}^N} h\,u\,d\mathbf{x}\right| \le \|h\|_{p/(p-1)}\|u\|_p.$$

Hence, from (G_p),

$$M \ge \Phi(u) \ge \frac{1}{2}\|\nabla u\|_2^2 + \mathbf{C}\|u\|_p^p - \|\varphi\|_1 - \|h\|_{p/(p-1)}\|u\|_p \tag{2.8}$$

holds for all $u \in E$. Denote

$$\|\nabla u\|_2^2 = s, \qquad \|u\|_p = t, \qquad \|h\|_{p/(p-1)} = A. \tag{2.9}$$

From the inequality

$$M \ge \frac{1}{2}|s|^2 + \mathbf{C}|t|^p - A|t| - \mathbf{C},$$

with positive constants A, $\mathbf{C}$, and an exponent $p > 1$, the boundedness of s and t follows, hence estimate (2.8) leads to the estimate

$$\|\nabla u\|_2 + \|u\|_p \le \mathbf{C}(M).$$

□

Boundedness of Φ from below. As in (2.8),

$$\Phi(u) \ge \mathbf{C}\|u\|_p^p - \|\varphi\|_1 - \|h\|_{p/(p-1)}\|u\|_p \tag{2.10}$$

for all $u \in \mathcal{E}_p$. Use the notation in (2.9). Since the function in t of the form

$$\mathbf{C}|t|^p - A|t| - \mathbf{C},$$

with some positive constants $\mathbf{C}$ and an exponent $p > 1$, is bounded from below, (2.10) yields the boundedness of Φ on $\mathcal{E}_p$. □

Weak lower semicontinuousness of Φ. Let

$$u_n \rightharpoonup u \qquad \text{in } \mathcal{E}_p. \tag{2.11}$$

From (2.11), we have

$$\lim_{n\to\infty} \int_{\mathbb{R}^N} h\,u_n\,d\mathbf{x} = \int_{\mathbb{R}^N} h\,u\,d\mathbf{x}. \tag{2.12}$$

Because $\{u_n\}$ satisfies (2.11),

$$\|\nabla u_n\|_2 + \|u_n\|_p \le \mathbf{C}. \tag{2.13}$$

On the other hand, condition (g_p) implies in accordance with the definition of K_p the existence of a constant $\mathbf{C}$ and a function $g_1 \in L_{p/(p-1)}(\mathbb{R}^N)$ such that

$$|g(\mathbf{x},t)| \le \mathbf{C}|t|^{p-1} + g_1(\mathbf{x})$$

holds for almost all $\mathbf{x} \in \mathbb{R}^N$ and for all $t \in \mathbb{R}$. Hence,

$$G(\mathbf{x},t) \le \int_0^t |g(\mathbf{x},s)|\,ds \le \mathbf{C}\,|t|^p + g_1(\mathbf{x})\,|t|. \tag{2.14}$$

From (2.14) and the Hölder inequality the estimate

$$\int_{\mathbb{R}^N} G(\mathbf{x},u_n)\,d\mathbf{x} \le \mathbf{C}\,\|u_n\|_p^p + \|g_1\|_{p/(p-1)}\|u_n\|_p \tag{2.15}$$

follows. Hence, from (2.13) and (2.15)

$$\int_{\mathbb{R}^N} G(\mathbf{x},u_n)\,d\mathbf{x} \le \mathbf{C}\,. \tag{2.16}$$

Taking into consideration that $\{u_n\} \subset \mathcal{E}_{2^*}(B_R)$ for any R and using the Kondrashov theorem, we get, by (2.11), that

$$u_n \to u \qquad \text{in } L_1(B_R) \tag{2.17}$$

for any B_R. The Lebesgue theory says that for any R one can choose from sequence (2.17) a subsequence convergent almost everywhere in B_R. Because B_R is arbitrary, we may choose a subsequence such that

$$u_n \to u \qquad \text{almost everywhere in } \mathbb{R}^N. \tag{2.18}$$

According to (G_p), we have for all n

$$G(\mathbf{x},u_n) \ge \varphi(\mathbf{x}) \in L_1(\mathbb{R}^N). \tag{2.19}$$

Properties (2.16), (2.18), and (2.19) allow the use of the Lebesgue-Fatou theorem and we get

$$\varliminf_{n\to\infty} \int_{\mathbb{R}^N} G(\mathbf{x},u_n)\,d\mathbf{x} \ge \int_{\mathbb{R}^N} G(\mathbf{x},u)\,d\mathbf{x}. \tag{2.20}$$

Lemma 2.4 provides

$$\varliminf_{n\to\infty} T(u_n) \ge T(u). \tag{2.21}$$

Combining (2.12), (2.20), (2.21), and taking into consideration that

$$\Phi(u_n) = \frac{1}{2}T(u_n) + \int_{\mathbb{R}^N} G(\mathbf{x},u_n)\,d\mathbf{x} - \int_{\mathbb{R}^N} h\,u_n\,d\mathbf{x},$$

we complete the proof. □

2.8 The Second Sample

In the first sample, conditions on functions g and h were strictly connected. We supposed that $h \in L_{p/(p-1)}(\mathbb{R}^N)$, where p was defined such that g belonged to K_p. This connection can be weakened provided $N > 2$. The particular formulation depends on how the exponents, which will be defined later, are related to the critical Sobolev exponent. This section is devoted to one of such variants. Others will be considered in the next sections.

THEOREM 2.7. *Let $N > 2$. Let conditions (g_q), (G_q), and (h_p) be satisfied for some $2^* \le p \le q < +\infty$. Then problem (P_1),*

$$-\Delta u + g(\mathbf{x}, u) = h(\mathbf{x}) \qquad \text{in } \mathbb{R}^N, \tag{P$_1$}$$

has a generalized solution $u \in \mathcal{E}_{p,q}$.

EXAMPLE. The problem

$$-\Delta u + \lambda |u|^{q-2} u = h(\mathbf{x}) \qquad \text{in } \mathbb{R}^N,$$

has a generalized solution $u \in \mathcal{E}_{p,q}$ for any $N > 2$, $2^* \le p \le q < +\infty$, $\lambda > 0$, and $h \in L_{p/(p-1)}(\mathbb{R}^N)$.

We shall not prove Theorem 2.7 separately, but formulate and prove a more general result, one of the corollaries of which is precisely Theorem 2.7. This generalization concerns in particular the behavior of g at both infinity and zero. Contrary to the conditions of Theorem 2.6 we allow the function g to behave in different ways at infinity and at zero. Thus, consider the problem

$$-\Delta u + g(\mathbf{x}, u) = h_1(\mathbf{x}) + h_2(\mathbf{x}) \qquad \text{in } \mathbb{R}^N, \tag{P$_2$}$$

where

$$g \in K_{p,q}, \tag{g$_{p\,q}$}$$

$$\begin{aligned} h_1 &\in L_{p/(p-1)}(\mathbb{R}^N), \\ h_2 &\in L_{q/(q-1)}(\mathbb{R}^N). \end{aligned} \tag{h$_{p\,q}$}$$

Here, as always, $1 < p \le q < +\infty$.

It is easy to see that in (P_2),

$$f \in K_{p,q} \qquad f(\mathbf{x}, t) = g(\mathbf{x}, t) - h_1(\mathbf{x}) - h_2(\mathbf{x}).$$

THEOREM 2.8. *Let conditions $(g_{p\,q})$, (G_q), and $(h_{p\,q})$ be satisfied for some $2^* \le p \le q < +\infty$. Then problem (P_2) has a generalized solution $u \in \mathcal{E}_{p,q}$.*

Remark. In accordance with the theorems of Section 1.2, if the functions g and h from Theorem 2.8 are regular enough, the generalized solution u is classical, and the boundary condition is also satisfied in the classical sense.

EXAMPLE. The problem

$$-\Delta u + |u|^{p-2}u(\lambda_1 + \lambda_2((\sin u)^2 + \frac{u}{p}\sin 2u)) + \lambda_3|u|^{q-2}u = h_1(\mathbf{x}) + h_2(\mathbf{x}) \text{ in } \mathbb{R}^N,$$

has a generalized solution $u \in \mathcal{E}_{p,q}$ for any $N > 2$, $2^* \le p \le q < +\infty$, $\lambda_1 \ge 0$, $\lambda_2 \ge 0$, $\lambda_3 > 0$, $h_1 \in L_{p/(p-1)}(\mathbb{R}^N)$, $h_2 \in L_{q/(q-1)}(\mathbb{R}^N)$.

Remark. Given $h \in L_s(\mathbb{R}^N)$ for some $1 \le s < +\infty$, then supposing

$$h_1 = \begin{cases} h, & |h| \ge 1, \\ 0, & |h| < 1, \end{cases}$$

we get $h = h_1 + h_2$, where $h_1 \in L_{s_1}(\mathbb{R}^N)$ for all $1 \le s_1 \le s$ and $h_2 \in L_{s_2}(\mathbb{R}^N)$ for all $s_2 \ge s$.

Proof of Theorem 2.8. Analogously to the proof of Theorem 2.6, consider the variational problem

$$\{u \in \mathcal{E}_{p,q}, \min \Phi(u)\}, \tag{V$_2$}$$

where the problem functional $\Phi : \mathcal{E}_{p,q} \to \mathbb{R}$ has the following form:

$$\begin{aligned} \Phi(u) &= \frac{1}{2}\|\nabla u\|_2^2 + V(u), \\ V(u) &= \int_{\mathbb{R}^N} G(\mathbf{x}, u)\, d\mathbf{x} - \int_{\mathbb{R}^N} h_1\, u\, d\mathbf{x} - \int_{\mathbb{R}^N} h_2\, u\, d\mathbf{x}. \end{aligned} \tag{2.22}$$

A solution of (V_2), if it exists, is a generalized solution of boundary problem (P_2) because of Theorem 2.1 and Theorem 1.6. To prove solvability of variational problem (V_2) we have in accordance with Theorem 2.5 to show that the functional Φ is

(*i*) bounded from below,
(*ii*) weakly lower semicontinuous,
(*iii*) coercive

on $\mathcal{E}_{p,q}$. We begin with the third condition.

Coerciveness of Φ. Choose M in such a way that the set

$$E \equiv \{u \in \mathcal{E}_{p,q}, |\Phi(u)| \le M\}$$

is nonempty. From condition (G_q) and the Hölder inequality we get analogously to (2.8) that

$$\begin{aligned} M \ge \Phi(u) \ge \frac{1}{2}\|\nabla u\|_2^2 + \mathbf{C}\,\|u\|_q^q - \|\varphi\|_1 \\ - \|h_1\|_{p/(p-1)}\|u\|_p - \|h_2\|_{q/(q-1)}\|u\|_q \end{aligned} \tag{2.23}$$

holds for all $u \in E$.

The Sobolev embedding theorem of Section 1.1 says that

$$\|u\|_p \leq \mathbf{C}\,(\|\nabla u\|_2 + \|u\|_q). \tag{2.24}$$

From (2.23), (2.24) we get

$$M \geq \frac{1}{2}\|\nabla u\|_2^2 + \|u\|_q^q - \mathbf{C}\,(\|\nabla u\|_2 + \|u\|_q) - \mathbf{C}. \tag{2.25}$$

Denote

$$\|\nabla u\|_2 = t, \qquad \|u\|_q = s.$$

Because

$$\mathbf{C}\,|t|^2 - \mathbf{C}\,|t| + \mathbf{C}\,|s|^q - \mathbf{C}\,|s| \leq \mathbf{C},$$

with positive constants $\mathbf{C}$ and $q > 1$, yields boundedness of t and s, estimate (2.25) implies

$$\|\nabla u\|_2 + \|u\|_q \leq \mathbf{C}(M). \tag{2.26}$$

(We recall that the same symbol $\mathbf{C}$ may denote different constants.) Then from (2.24) and (2.26) it follows that

$$\|\nabla u\|_2 + \|u\|_p + \|u\|_q \leq \mathbf{C}(M). \qquad \square$$

Boundedness of Φ from below. Analogously to (2.25)

$$\Phi(u) \geq \frac{1}{2}\|\nabla u\|_2^2 + \|u\|_q^q - \mathbf{C}\,(\|\nabla u\|_2 + \|u\|_q) \geq -\mathbf{C}. \qquad \square$$

Weak lower semicontinuousness of Φ. Let

$$u_n \rightharpoonup u \qquad \text{in } \mathcal{E}_{p,q}. \tag{2.27}$$

From (2.27) we have

$$\begin{aligned} \lim_{n\to\infty} \int_{\mathbb{R}^N} h_1\, u_n\, d\mathbf{x} &= \int_{\mathbb{R}^N} h_1\, u\, d\mathbf{x}, \\ \lim_{n\to\infty} \int_{\mathbb{R}^N} h_2\, u_n\, d\mathbf{x} &= \int_{\mathbb{R}^N} h_2\, u\, d\mathbf{x}. \end{aligned} \tag{2.28}$$

Because u_n satisfies (2.27),

$$\|\nabla u_n\|_2 + \|u_n\|_p + \|u_n\|_q \leq \mathbf{C}. \tag{2.29}$$

On the other hand, from condition $(g_{p\,q})$ and the Hölder inequality we get analogously to (2.14) that

$$\int_{\mathbb{R}^N} G(\mathbf{x}, u_n)\, d\mathbf{x} \leq \mathbf{C}\,(\|u_n\|_p^p + \|u_n\|_q^q) \\ + \|g_1\|_{p/(p-1)}\|u_n\|_p + \|g_2\|_{q/(q-1)}\|u_n\|_q, \tag{2.30}$$

where g_1 and g_2 are defined by g belonging to $K_{p,q}$ in accordance with the definition of $K_{p,q}$ in Section 1.3. Consequently, by (2.29) and (2.30),

$$\int_{\mathbb{R}^N} G(\mathbf{x}, u_n)\, d\mathbf{x} \leq \mathbf{C}. \tag{2.31}$$

Taking into consideration the Kondrashov theorem we get, by (2.27), that

$$u_n \to u \qquad \text{in } L_1(B_R)$$

for any B_R.

Analogously to the inference (2.17) $\Rightarrow$ (2.18), we have up to a subsequence

$$u_n \to u \qquad \text{almost everywhere in } \mathbb{R}^N. \tag{2.32}$$

Condition (G_q) gives

$$G(u_n) \geq \varphi(\mathbf{x}), \qquad \varphi \in L_1(\mathbb{R}^N). \tag{2.33}$$

Properties (2.31), (2.32), and (2.33) allow us to use the Lebesgue-Fatou theorem. Hence, we get

$$\varliminf_{n\to\infty} \int_{\mathbb{R}^N} G(\mathbf{x}, u_n)\, d\mathbf{x} \geq \int_{\mathbb{R}^N} G(\mathbf{x}, u)\, d\mathbf{x}. \tag{2.34}$$

Recall that from Lemma 2.4,

$$\varliminf_{n\to\infty} T(u_n) \geq T(u). \tag{2.35}$$

Combining (2.28), (2.34), and (2.35) and taking into consideration that

$$\Phi(u_n) = \frac{1}{2}T(u_n) + \int_{\mathbb{R}^N} G(\mathbf{x}, u_n)\, d\mathbf{x} - \int_{\mathbb{R}^N} h_1\, u_n\, d\mathbf{x} - \int_{\mathbb{R}^N} h_2\, u_n\, d\mathbf{x},$$

we complete the proof. □

2.9 THE THIRD SAMPLE

Here we consider problems (P_1), (P_2) with p and q in another relation to the critical exponent 2^*. More precisely, let $p \le 2^*$, $q \le 2^*$. In full analogy to Theorems 2.7 and 2.8, we formulate two new theorems.

THEOREM 2.9. *Let $N > 2$. Let conditions (g_p), (G_p), and (h_q) be satisfied with some $1 < p \le q \le 2^*$. Then problem (P_1),*

$$-\Delta u + g(\mathbf{x}, u) = h(\mathbf{x}) \qquad \text{in } \mathbb{R}^N, \tag{P$_1$}$$

has a generalized solution $u \in \mathcal{E}_{p,q}$.

EXAMPLE. The problem

$$-\Delta u + \lambda |u|^{p-2} u = h(\mathbf{x}) \qquad \text{in } \mathbb{R}^N,$$

has a generalized solution $u \in \mathcal{E}_{p,q}$ for any $N > 2$, $1 < p \le q \le 2^*$, $\lambda > 0$, and $h \in L_{q/(q-1)}(\mathbb{R}^N)$.

THEOREM 2.10. *Let $N > 2$. Let conditions $(g_{p\,q})$, (G_p) and $(h_{p\,q})$ be satisfied for some $1 < p \le q \le 2^*$. Then the problem*

$$-\Delta u + g(\mathbf{x}, u) = h_1(\mathbf{x}) + h_2(\mathbf{x}) \qquad \text{on } \mathbb{R}^N, \tag{P$_2$}$$

has a generalized solution $u \in \mathcal{E}_{p,q}$.

Remark. In accordance with the theorems of Section 1.2, if the functions g and h from Theorem 2.10 are regular enough, the generalized solution u is classical and the boundary condition is also satisfied in the classical sense.

EXAMPLE. The problem

$$-\Delta u + \lambda_1 |u|^{p-2} u + |u|^{q-2} u (\lambda_2 + \lambda_3 ((\sin u)^2 + \frac{u}{q} \sin 2u)) = h_1(\mathbf{x}) + h_2(\mathbf{x}) \,\text{on}\, \mathbb{R}^N,$$

has a generalized solution $u \in \mathcal{E}_{p,q}$ for any $N > 2$, $1 < p \le q \le 2^*$, $\lambda_1 > 0$, $\lambda_2 \ge 0$, $\lambda_3 \ge 0$, $h_1 \in L_{p/(p-1)}(\mathbb{R}^N)$, and $h_2 \in L_{q/(q-1)}(\mathbb{R}^N)$.

Theorem 2.9 is a consequence of Theorem 2.10, analogously to Section 2.4.

Proof of Theorem 2.9. As was done in previous sections, consider the variational problem

$$\{u \in \mathcal{E}_{p,q}, \min \Phi(u)\}, \tag{V$_2$}$$

where the problem functional $\Phi : \mathcal{E}_{p,q} \to \mathbb{R}$ has the form of (2.22).

A solution of (V_2), if such exists, is a generalized solution of boundary problem (P_2) from Theorem 2.1 and Theorem 1.6. To prove the solvability of variational problem (V_2), we need in accordance with Theorem 2.5 to show that the functional Φ is

(*i*) bounded from below,
(*ii*) weakly lower semicontinuous,
(*iii*) coercive

on $\mathcal{E}_{p,q}$. We begin with the third property.

Coerciveness of Φ. Choose M in such a way that the set

$$E \equiv \{u \in \mathcal{E}_{p,q}, |\Phi(u)| \leq M\}$$

is nonempty. From condition (G_p) and the Hölder inequality we get that

$$\begin{aligned} M \geq \Phi(u) \geq \frac{1}{2}\|\nabla u\|_2^2 + \mathbf{C}\,\|u\|_p^p - \|\varphi\|_1 \\ - \|h_1\|_{p/(p-1)}\|u\|_p - \|h_2\|_{q/(q-1)}\|u\|_q \end{aligned} \tag{2.36}$$

holds for all $u \in E$.

The Sobolev embedding theorem of Section 1.1 says that

$$\|u\|_q \leq \mathbf{C}\,(\|\nabla u\|_2 + \|u\|_p). \tag{2.37}$$

From (2.36) and (2.37) we get

$$M \geq \frac{1}{2}\|\nabla u\|_2^2 + \mathbf{C}\,\|u\|_p^p - \mathbf{C}\,(\|\nabla u\|_2 + \|u\|_p) - \mathbf{C}. \tag{2.38}$$

Analogously to (2.26), estimate (2.38) implies

$$\|\nabla u\|_2 + \|u\|_p \leq \mathbf{C}(M). \tag{2.39}$$

Combining (2.37) and (2.39) gives the required result. □

Boundedness of Φ may be proved as in the previous section. □

Weak lower semicontinuousness of Φ may be proved as in the previous section. □

2.10 THE FOURTH SAMPLE

In this section, we expand as compared with the second and third examples the domains of definition for p, q, and N. But we pay for this by a condition on g stronger than conditions (G_p) or (G_q).

Thus, suppose that there exist positive constants C_1, C_2, and a function $\varphi \in L_1(\mathbb{R}^N)$ such that

$$G(\mathbf{x}, t) \geq C_1 |t|^p + C_2 |t|^q + \varphi(\mathbf{x}) \qquad (\mathrm{G}_{p\,q})$$

for almost all $\mathbf{x} \in \mathbb{R}^N$ and for all $t \in \mathbb{R}$.

THEOREM 2.11. *Let conditions* $(g_{p\,q})$, $(G_{p\,q})$, *and* $(h_{p\,q})$ *be satisfied with some* $1 < p \leq q < \infty$. *Then problem* (P_1),

$$-\Delta u + g(\mathbf{x}, u) = h(\mathbf{x}) \qquad \text{in } \mathbb{R}^N, \qquad (\mathrm{P}_1)$$

has a generalized solution $u \in \mathcal{E}_{p,q}$.

Remark. In accordance with the theorems of Section 1.2, if the functions g and h from Theorem 2.11 are regular enough, the generalized solution u is classical and the boundary condition is also satisfied in the classical sense.

EXAMPLE. The problem

$$-\Delta u + \lambda_1 |u|^{p-2} u + \lambda_2 |u|^{q-2} u = h_1(\mathbf{x}) + h_2(\mathbf{x}) \text{ in } \mathbb{R}^N,$$

has a generalized solution $u \in \mathcal{E}_{p,q}$ for any N, $1 < p \leq q < +\infty$, $\lambda_1 > 0$, $\lambda_2 > 0$, $h \in L_{q/(q-1)}(\mathbb{R}^N)$, and $h \in L_{p/(p-1)}(\mathbb{R}^N)$.

Proof of Theorem 2.11. As in the previous sections, consider the variational problem

$$\{u \in \mathcal{E}_{p,q}, \min \Phi(u)\}, \qquad (\mathrm{V}_2)$$

where the problem functional $\Phi : \mathcal{E}_{p,q} \to \mathbb{R}$ has the form of (2.22). A solution of (V_2), if it exists, is a generalized solution of boundary problem (P_2) from Theorem 2.1 and Theorem 1.6. To prove the solvability of variational problem (V_2) we need in accordance with Theorem 2.5 to show that the functional Φ is

(i) bounded from below,
(ii) weakly lower semicontinuous,
(iii) coercive

on $\mathcal{E}_{p,q}$.

Coerciveness of Φ. Choose M in such a way that the set

$$E \equiv \{u \in \mathcal{E}_{p,q}, |\Phi(u)| \leq M\}$$

is nonempty. From condition (G_{pq}) and the Hölder inequality we get

$$\begin{aligned} M \geq \Phi(u) \geq \frac{1}{2}\|\nabla u\|_2^2 + \mathbf{C}\left(\|u\|_p^p + \|u\|_q^q\right) - \|\varphi\|_1 \\ - \|h_1\|_{p/(p-1)}\|u\|_p - \|h_2\|_{q/(q-1)}\|u\|_q \end{aligned} \tag{2.40}$$

for all $u \in E$. As was done previously, inequality (2.40), with $p > 1$, $q > 1$, causes

$$\|\nabla u\|_2 + \|u\|_p + \|u\|_q \leq \mathbf{C}(M). \qquad \square$$

Boundedness of Φ *from below.* From condition (G_{pq}) and the Hölder inequality

$$\Phi(u) \geq \mathbf{C}\left(\|u\|_p^p + \|u\|_q^q\right) - \|\varphi\|_1 - \|h_1\|_{p/(p-1)}\|u\|_p - \|h_2\|_{q/(q-1)}\|u\|_q. \tag{2.41}$$

It is easy to see that the left-hand side of (2.41) is bounded from below, and so is Φ.
$\square$

Weak lower semicontinuousness of Φ may be proved as in Section 2.8. $\square$

§3. Approximation by Bounded Domains

Direct variational methods are not always applicable to problems of the form

$$-\Delta u + f(\mathbf{x}, u) = 0 \qquad \text{in } \mathbb{R}^N, \tag{3.1}$$

especially in the case where a problem functional with "good" properties can not be defined. However, one often is successful in approximating a solution to the problem by solutions of equation (3.1) considered, for instance, in the open balls

$$\begin{aligned} -\Delta u + f(\mathbf{x}, u) &= 0 \qquad \text{in } B_R, \\ u(\partial B_R) &= 0. \end{aligned} \tag{3.2}$$

The corresponding problem functionals possessing the necessary properties can be defined on B_R, which permits us to study problem (3.2) by means of known variational methods. A solution to (3.1) may be then obtained by the limit procedure as $R \to +\infty$.

3.1 GENERALIZED SOLUTION IN THE CASE OF BALLS

Because we will approximate the solutions in $\mathbb{R}^N$ by solutions in balls B_R with finite R, we introduce corresponding functional spaces for the functions defined on these balls.

Define the spaces $\mathcal{E}_p^0(B_R)$, $1 \le p < +\infty$, as closures of $\mathcal{D}(B_R)$ in the corresponding norms

$$\|\nabla u\|_2(B_R) + \|u\|_p(B_R)$$

analogously to how $\mathcal{E}_p$ was defined for $\mathbb{R}^N$. Then, in particular, in accordance with the definition of Section 1.1, $H_0^1(B_R) = \mathcal{E}_2^0(B_R)$. For bounded domains the corresponding embedding theorems are true.

SOBOLEV EMBEDDING THEOREMS FOR BALLS.

$$H_0^1(B_R) \circlearrowright L_p(B_R) \qquad \text{if } 1 \le p \le 2^*,$$
$$H_0^1(B_R) \circlearrowright \mathcal{E}_p^0(B_R) \circlearrowright H_0^1(B_R) \qquad \text{if } 1 \le p \le 2^*.$$

Hence, in particular, for any $1 \le p \le 2^*$ there is a constant $C_p > 0$ such that for all $u \in H_0^1(B_R)$

$$\|\nabla u\|_2 \ge C_p \|u\|_p.$$

The spaces $\mathcal{E}_p(B_R)$ can be defined as the closure of infinitely differentiable functions $u \in C^\infty(B_R)$ in the same norm as for $\mathcal{E}_p^0(B_R)$. Note that $\mathcal{E}_p^0(B_R) \neq \mathcal{E}_p(B_R)$ for finite R. Spaces $\mathcal{E}_p(B_R)$ may also be considered as the spaces of functions $u \in \mathcal{E}_p(\mathbb{R}^N)$ reduced onto B_R.

Note that if $u \in \mathcal{E}_p^0(B_R)$, then $u(\mathbf{x}) = 0$ for $\mathbf{x} \in \partial B_R$ in a sense. Therefore, in the generalized setting of the problems in balls, the Dirichlet condition $u(\mathbf{x}) = 0$ for $\mathbf{x} \in \partial B_R$ may actually be included in the condition $u \in \mathcal{E}_p^0(B_R)$ for corresponding p.

Indeed, generalized solutions to (3.2) may be defined analogously to the procedure in Section 1.2 in the following way:

DEFINITION. A function $u \in \mathcal{E}_p^0(B_R)$ is a generalized solution of (3.2) if the function $f(\,.\,, u(\,.\,))$ is locally integrable and for all $\varphi \in \mathcal{D}(B_R)$,

$$\int_{B_R} (\nabla u, \nabla \varphi)\, d\mathbf{x} + \int_{B_R} f(\mathbf{x}, u)\varphi\, d\mathbf{x} = 0.$$

DEFINITION. A function $u \in C^2(B_R) \cap C(\bar{B}_R)$ is a classical solution to (3.2) if after substituting it into equation (3.2), this equation becomes the identity at each point $\mathbf{x} \in B_R$, and $u(\mathbf{x}) = 0$ provided $\mathbf{x} \in \partial B_R$.

For generalized solutions, the regularity theorems analogous to the theorems of Section 1.2 are true. They differ only by the replacement of $\mathbb{R}^N$ by B_R. This means that if f is regular enough, then any generalized solution to (3.2) is a classical solution.

Let $f \in K_p(B_R)$ for some $1 < p < +\infty$. Analogously to Section 1.3, the problem functional $\Phi : \mathcal{E}_p^0(B_R) \to \mathbb{R}$ of the class C^1,

$$\Phi(u) = \frac{1}{2}\|\nabla u\|_2^2 + \int\limits_{B_R} F(\mathbf{x}, u)\, d\mathbf{x}, \tag{3.3}$$

with

$$F(\mathbf{x}, t) = \int\limits_0^t f(\mathbf{x}, s)\, ds,$$

can be defined.

The variational method for such domains is based on the following theorem analogous to Theorem 1.6.

Theorem 3.1. *Let $f \in K_p(B_R)$, and $u \in \mathcal{E}_p^0(B_R)$ be a critical point of the problem functional $\Phi : \mathcal{E}_p^0(B_R) \to \mathbb{R}$. Then u is a generalized solution of (3.2).*

Furthermore, any function $u \in \mathcal{E}_p^0(B_R)$ can be extended onto $\mathbb{R}^N$ by

$$\widetilde{u}(\mathbf{x}) = \begin{cases} u(\mathbf{x}), & \mathbf{x} \in B_R, \\ 0, & \mathbf{x} \in \mathbb{R}^N \setminus B_R. \end{cases}$$

By this extension we get $\widetilde{u} \in \mathcal{E}_p$ and analogously $\widetilde{u} \in \mathcal{E}_p^0(B_{R_0})$ for all $R_0 \geq R$ (Gilbarg and Trudinger [1]). We shall identify u and $\widetilde{u}$ and consider $u \in \mathcal{E}_p^0(B_R)$ as an element of $\mathcal{E}_p$.

Remark. The definitions of this section can be generalized to the case of smooth bounded domains $\Omega \in \mathbb{R}^N$.

3.2 The Theorem on Approximation by Bounded Domains

The method of approximation by bounded domains is based on the following theorem.

Theorem 3.2. *Let a function f be represented in the form $f(\mathbf{x}, t) = g(\mathbf{x}, t) - h(\mathbf{x})$, where $g \in K_{p_0}(B_R)$ with $2 < p_0 < 2^* + 1$ for any R, and $h \in L_1(B_R)$ also for any R.*

Let for some $1 < p < +\infty$ there be a sequence of balls B_{R_n}, with $R_n \to \infty$ as $n \to \infty$, such that each problem

$$\begin{aligned} &-\Delta u + f(\mathbf{x}, u) = 0 \qquad \text{in } B_{R_n}, \\ &u(\mathbf{x}) = 0 \qquad \text{on } \partial B_{R_n}, \end{aligned} \tag{P$_n$}$$

possesses a solution $u_n \in \mathcal{E}_p^0(B_{R_n})$ *such that*

$$\|\nabla u_n\|_2(B_{R_n}) + \|u_n\|_p(B_{R_n}) \leq \mathbf{C}, \tag{3.4}$$

where the constant $\mathbf{C}$ *does not depend on* n.

Then the problem

$$-\Delta u + f(\mathbf{x}, u) = 0 \qquad \text{in } \mathbb{R}^N, \tag{P$_\infty$}$$

has a generalized solution $u \in \mathcal{E}_p$, *and there is a subsequence* $\{u_n\}$ *such that*

$$u_n \rightharpoonup u \qquad \text{in } \mathcal{E}_p \qquad \text{as } n \to \infty.$$

Before starting the proof we present one well-known fact from the theory of nonlinear operators which is used in the given proof and which is referred to later on.

3.3 THE VAINBERG-KRASNOSELSKII THEOREM

Let $f : \Omega \times \mathbb{R} \to \mathbb{R}$ be a Carathéodory function, and let there exist a constant $\mathbf{C} > 0$, exponents $1 \leq p \leq q < +\infty$, and a function $g \in L_{q/p}(\Omega)$ such that the inequality

$$|f(\mathbf{x}, t)| \leq g(\mathbf{x}) + \mathbf{C}\,|t|^p$$

holds for almost all $\mathbf{x} \in \Omega$ and for all $t \in \mathbb{R}$.

THEOREM. *Let the above assumptions hold and* $\{u_n\}$ *be a sequence from* $L_q(\Omega)$ *such that*

$$u_n \to u \qquad \text{in } L_q(\Omega)$$

as $n \to +\infty$. *Then*

$$f(\,.\,, u_n(\,.\,)) \to f(\,.\,, u(\,.\,)) \qquad \text{in } L_{q/p}(\Omega).$$

Remark. By the Hölder inequality, it follows that if Ω is a bounded domain,

$$f(\,.\,, u_n(\,.\,)) \to f(\,.\,, u(\,.\,)) \qquad \text{in } L_r(\Omega)$$

for all $1 \leq r \leq q/p$.

Proof. See Vainberg [1] or Krasnoselskii [1]. □

3.4 THE PROOF OF THE THEOREM ON APPROXIMATION BY BOUNDED DOMAINS

Now we are ready to prove Theorem 3.2.

Because of the reflexivity of $\mathcal{E}_p$ and condition (3.4) we may choose a subsequence of the sequence $\{u_n\}$ such that

$$u_n \rightharpoonup u \qquad \text{in } \mathcal{E}_p. \tag{3.5}$$

We shall show that u is a generalized solution to (P_∞). Choose an arbitrary $\varphi \in \mathcal{D}(\mathbb{R}^N)$. According to the definition of $\mathcal{D}(\mathbb{R}^N)$, the support of φ is bounded in $\mathbb{R}^N$, that is, the existence of a ball B_{R_0} such that $\operatorname{supp}\varphi \subset B_{R_0}$ follows.

Because u_n is a generalized solution to (P_n), then

$$\begin{aligned}\int_{\mathbb{R}^N} (\nabla u_n, \nabla\varphi)\,d\mathbf{x} &= \int_{B_{R_n}} (\nabla u_n, \nabla\varphi)\,d\mathbf{x} \\ &= -\int_{B_{R_n}} f(\mathbf{x}, u_n)\varphi\,d\mathbf{x} = -\int_{B_{R_0}} f(\mathbf{x}, u_n)\varphi\,d\mathbf{x}\end{aligned} \tag{3.6}$$

holds for all $R_n > R_0$. Owing to weak convergence (3.5), we have the following limit relation for the left-hand side of (3.6):

$$\lim_{n\to\infty} \int_{\mathbb{R}^N} (\nabla u_n, \nabla\varphi)\,d\mathbf{x} = \int_{\mathbb{R}^N} (\nabla u, \nabla\varphi)\,d\mathbf{x}.$$

On the other hand, from the Sobolev and Kondrashov theorems and (3.5)

$$u_n \to u \qquad \text{in } L_{p_0-1}(B_{R_0}) \qquad \text{as } n \to \infty.$$

Because $g \in K_{p_0}(B_{R_0})$, the Vainberg-Krasnoselskii theorem gives

$$\varphi f(.,u_n(.)) \to \varphi f(.,u(.)) \qquad \text{in } L_1(B_{R_0}) \qquad \text{as } n \to \infty.$$

Then the limit relation for the right-hand side of (3.6) follows:

$$\lim_{n\to\infty} \int_{B_{R_0}} f(\mathbf{x}, u_n)\varphi\,d\mathbf{x} = \int_{B_{R_0}} f(\mathbf{x}, u)\varphi\,d\mathbf{x}.$$

Thus, passing to the limit in (3.6), we get

$$\int_{\mathbb{R}^N} (\nabla u, \nabla\varphi)\,d\mathbf{x} = -\int_{B_{R_0}} f(\mathbf{x}, u)\varphi\,d\mathbf{x} = -\int_{\mathbb{R}^N} f(\mathbf{x}, u)\varphi\,d\mathbf{x},$$

which corresponds to the definition of a weak solution. It is generalized by the force of (3.5). □

Some applications of this theorem are presented in the next sections.

3.5 The Theorem on Approximation of Bounded Domains for Eigenvalues Problems

One may propose a modification of Theorem 3.2 for a problem with a parameter λ of the following form:

$$-\Delta u + f_1(\mathbf{x}, u) + \lambda f_2(\mathbf{x}, u) = 0 \qquad \text{in } \mathbb{R}^N. \tag{P$_\lambda$}$$

THEOREM 3.3. *Let functions f_i with $i = 1$ or 2 be represented in the form $f_i(\mathbf{x}, t) = g_i(\mathbf{x}, t) - h_i(\mathbf{x})$, where $g_i \in K_{p_i}(B_R)$, with $2 < p_i < 2^* + 1$ for any R, and $h_i \in L_1(B_R)$ also for any R.*

Let for some $1 < p < +\infty$ there be a sequence of balls B_{R_n}, with $R_n \to \infty$ as $n \to \infty$, and a sequence of numbers $\{\lambda_n\}$ such that each problem

$$\begin{aligned} &-\Delta u + f_1(\mathbf{x}, u) + \lambda_n f_2(\mathbf{x}, u) = 0 \qquad \text{in } \mathbb{R}^N, \\ &u(\partial B_{R_n}) = 0, \end{aligned} \tag{P$_{\lambda,n}$}$$

possesses a solution $u_n \in \mathcal{E}_p^0(B_{R_n})$ such that

$$\begin{aligned} &\|\nabla u_n\|_2(B_{R_n}) + \|u_n\|_p(B_{R_n}) \le \mathbf{C}_1, \\ &|\lambda_n| \le \mathbf{C}_2, \end{aligned} \tag{3.7}$$

where constants $\mathbf{C}_i$ do not depend on n.

Then there exists a $\lambda \in \mathbb{R}$ such that the problem $(\mathrm{P}_{\lambda,\infty})$ has a solution $u \in \mathcal{E}_p$, and there are corresponding subsequences such that

$$\begin{aligned} u_n &\rightharpoonup u \qquad \text{in } \mathcal{E}_p, \\ \lambda_n &\to \lambda \qquad \text{in } \mathbb{R} \end{aligned}$$

as $n \to \infty$.

Proof. Because of the boundedness of $\{\lambda_n\}$ in $\mathbb{R}$, we may choose a subsequence such that

$$\lambda_n \to \lambda \qquad \text{in } \mathbb{R}.$$

The rest of the proof is analogous to the proof of Theorem 3.2, and we omit it. □

The results of the following section will be used in the next chapters.

§4. Approximation for Problems on an Absolute Minimum

In this section, a variant of this general method will be applied to problems that do not satisfy the conditions of the existence theorems of Sections 2.5–2.10.

4.1 THE MAIN RESULT

Let us consider the problem

$$-\Delta u + g(\mathbf{x}, u) = h(\mathbf{x}) \qquad \text{in } \mathbb{R}^N, \tag{P$_*$}$$

where

$$h \in L_{2^*/(2^*-1)}(\mathbb{R}^N). \tag{h$_*$}$$

Suppose that there exists a function $\varphi \in L_1(\mathbb{R}^N)$ such that

$$G(\mathbf{x}, t) \geq \varphi(\mathbf{x}) \tag{G$_*$}$$

for almost all $\mathbf{x} \in \mathbb{R}^N$ and for all $t \in \mathbb{R}$. The following theorem is true:

THEOREM 4.1. *Let $N > 2$ and conditions (G_*), (h_*) be satisfied. Let $g \in K_{2^*}(B_R)$ for any R. Then problem (P_*) has a generalized solution $u \in \mathcal{E}_{2^*}$.*

Proof. We want to make use of Theorem 3.2. For this purpose consider a family of problems

$$\begin{aligned} -\Delta u + f(\mathbf{x}, u) &= 0 \qquad \text{in } B_R, \\ u(\partial B_R) &= 0, \end{aligned} \tag{P$_R$}$$

and show that problem (P_R) has a solution $u_R \in \mathcal{E}_{2^*}$ for all R, the following estimate being true:

$$\|\nabla u_R\|_2 + \|u_R\|_{2^*} \leq \mathbf{C}, \tag{4.1}$$

with $\mathbf{C}$ not depending on R. □

First, we study problem (P_R) with respect to the existence of a solution.

4.2 THE EXISTENCE OF A SOLUTION IN BALLS

For proving the solvability of (P_R) with some fixed R consider the variational problem

$$\left\{u \in \mathcal{E}^0_{2^*}(B_R), \min \Phi_R(u)\right\}, \tag{V$_R$}$$

where the problem functional $\Phi_R : \mathcal{E}^0_{2^*}(B_R) \to \mathbb{R}$ is defined as

$$\Phi_R(u) = \frac{1}{2}\|\nabla u\|_2^2(B_R) + \int_{B_R} G(\mathbf{x}, u)\, d\mathbf{x} - \int_{B_R} h\, u\, d\mathbf{x}.$$

A solution of (V_R), if it exists, is a solution of boundary problem (P_R) from Theorem 2.1 and Theorem 1.6. To prove the solvability of variational problem (V_R), we need in accordance with Theorem 2.5 to show that the problem functional Φ is

(i) bounded from below,
(ii) weakly lower semicontinuous,
(iii) coercive

on $\mathcal{E}_{2^*}(B_R)$. Proving these properties, we ensure the existence of a solution to (P_R). We begin with the third property.

Coerciveness of Φ_R. Choose M in such a way that the set

$$E_R \equiv \left\{u \in \mathcal{E}^0_{2^*}(B_R), |\Phi_R(u)| \leq M\right\} \tag{4.2}$$

is nonempty. Then from condition (G_*) and the Hölder inequality, the inequality

$$M \geq \Phi_R(u) \geq \frac{1}{2}\|\nabla u\|_2^2 - \|\varphi\|_1 - \|h\|_{2^*/(2^*-1)}\|u\|_{2^*} \tag{4.3}$$

holds for all $u \in E_R$. In accordance with the Sobolev embedding theorem, (4.3) yields

$$M \geq \mathbf{C}\,\|u\|_{2^*}^2 - \|h\|_{2^*/(2^*-1)}\|u\|_{2^*} - \mathbf{C}\,, \tag{4.4}$$

and therefore

$$\|u\|_{2^*} \leq \mathbf{C}(M). \tag{4.5}$$

Estimates (4.3) and (4.5) imply

$$\|\nabla u\|_2 + \|u\|_{2^*} \leq \mathbf{C}(M).$$

□

Boundedness of Φ_R *from below.* From conditions (G_*), (h_*), the Hölder inequality, and the Sobolev embedding theorem,

$$\begin{aligned}\Phi_R(u) &\geq \frac{1}{2}\|\nabla u\|_2^2 - \|\varphi\|_1 - \|h\|_{2^*/(2^*-1)}\|u\|_{2^*}\\ &\geq \mathbf{C}\,\|u\|_{2^*}^2 - \|\varphi\|_1 - \|h\|_{2^*/(2^*-1)}\|u\|_{2^*} \geq -\mathbf{C}\end{aligned}$$

for all $u \in \mathcal{E}^0_{2^*}(B_R)$. □

Weak lower semicontinuousness of Φ_R. Let

$$u_n \rightharpoonup u \qquad \text{in } \mathcal{E}^0_{2^*}(B_R). \tag{4.6}$$

From (4.6) we have

$$\lim_{n\to\infty}\int_{B_R} h\,u_n\,d\mathbf{x} = \int_{B_R} h\,u\,d\mathbf{x}. \tag{4.7}$$

From $g \in K_{2^*}(B_R)$,

$$\int_{B_R} G(\mathbf{x}, u_n)\,d\mathbf{x} \leq \mathbf{C}\,\|u_n\|_{2^*}^{2^*} + \mathbf{C}. \tag{4.8}$$

From (4.6), (4.8), condition (G$_*$), and the Lebesgue-Fatou theorem,

$$\underline{\lim}_{n\to\infty} \int\limits_{B_R} G(\mathbf{x}, u_n)\, d\mathbf{x} \geq \int\limits_{B_R} G(\mathbf{x}, u)\, d\mathbf{x}. \tag{4.9}$$

Finally, as in Lemma 2.4,

$$\underline{\lim}_{R\to\infty} \|\nabla u_n\|_2^2(B_R) \geq \|\nabla u\|_2^2(B_R). \tag{4.10}$$

Combining (4.7), (4.9), and (4.10), and taking into consideration the definition of Φ_R, we complete the proof.

Thus, the existence of a solution $u_R \in \overset{\circ}{H}{}^1_0(B_R)$ to (V_R) and, consequently, to (P_R) is proved. □

4.3 Completing the Proof of Theorem 4.1: Boundedness of a Sequence

In order to finish the proof of Theorem 4.1, we need only to prove estimate (4.1).

Note that because of the definition, $\Phi_R(0) = 0$, and consequently,

$$\Phi_R(u_R) = \min_{u\in H_0^1(B_R)} \Phi_R(u_R) \leq 0 \tag{4.11}$$

for any R. From (4.11), estimate (G$_*$), and the Hölder inequality,

$$0 \geq \frac{1}{2}\|\nabla u_R\|_2^2 - \|\varphi\|_1 - \|h\|_{2^*/(2^*-1)}\|u_R\|_{2^*}. \tag{4.12}$$

From (4.12) and the Sobolev embedding theorem

$$0 \geq \mathbf{C}\,\|u_R\|_{2^*}^2 - \|\varphi\|_1 - \|h\|_{2^*/(2^*-1)}\|u_R\|_{2^*},$$

where $\mathbf{C}$ does not depend on R, and consequently,

$$\|u_R\|_{2^*} \leq \mathbf{C}. \tag{4.13}$$

Estimates (4.12) and (4.13) imply (4.1). Thus, Theorem 4.1 is proved. □

Example. The problem

$$-\Delta u + \lambda|u|^{p-2}u((\sin u)^2 + \frac{u}{p}\sin 2u) = h(\mathbf{x}) \qquad \text{in } \mathbb{R}^N,$$

has a generalized solution $u \in \mathcal{E}_{2^*}$ for any $N > 2$, $1 < p \leq 2^*$, $\lambda \geq 0$, and $h \in L_{2^*/(2^*-1)}$.

§5. The Monotonicity Method. Uniqueness of Solutions

In the study of various differential equations, the monotonicity method proves very helpful. Actually, it was used in Section 2.3 where the weak lower semicontinuousness of T was proved. However, because of the relative simplicity of problems investigated in this book, the existence results obtained by this method are as a rule weaker than those obtained by other methods.

On the other hand, the uniqueness theorem presented below requires the monotonicity approach.

More detailed information on the method itself and its applications to other problems may be found, for instance, in Vainberg [2].

5.1 A Uniqueness Theorem

Consider the following problem:

$$-\Delta u + g(\mathbf{x}, u) = h(\mathbf{x}) \qquad \text{on } \mathbb{R}^N.$$

THEOREM 5.1. *Let $h = h_1 + h_2$, $h_1 \in L_p(\mathbb{R}^N)$, $h_2 \in L_q(\mathbb{R}^N)$ for some $1 < p \leq q < +\infty$. Let the function $g \in K_{p,q}$ be nondecreasing (monotonic), that is,*

$$g(\mathbf{x}, t_1) - g(\mathbf{x}, t_2) \geq 0$$

for almost all $\mathbf{x} \in \mathbb{R}^N$ and for all $t_1 \geq t_2$ in $\mathbb{R}$. Then problem (5.1) has at most one generalized solution in $\mathcal{E}_{p,q}$.

Proof. As Theorem 1.7 asserts, any solution $u \in \mathcal{E}_{p,q}$ is a critical point of the problem functional $\Phi : \mathcal{E}_{p,q} \to \mathbb{R}$. Consequently,

$$\langle \Phi'(u), \varphi \rangle = 0, \tag{5.2}$$

$$\langle \Phi'(v), \varphi \rangle = 0 \tag{5.3}$$

for any two solutions $u, v \in \mathcal{E}_{p,q}$ and any function $\varphi \in \mathcal{E}_{p,q}$, where (5.2) and (5.3) are considered in accordance with Section 1.3. Choose $\varphi = u - v$ and subtract (5.3) from (5.2). Then

$$0 = \langle \Phi'(u) - \Phi'(v), u - v \rangle = \|\nabla(u - v)\|_2^2 + \int_{\mathbb{R}^N} (g(\mathbf{x}, u) - g(\mathbf{x}, v))(u - v)\, d\mathbf{x} \geq \|\nabla(u - v)\|_2^2. \tag{5.4}$$

It follows from (5.4) that

$$\|\nabla(u - v)\|_2^2 = 0$$

for any two solutions $u, v \in \mathcal{E}_{p,q}$. Hence, there exists a constant A such that

$$u = v + A. \tag{5.5}$$

Because u, v lie in $L_p(\mathbb{R}^N)$, then $A = 0$, that is, $u = v$. $\square$

5.2 EXAMPLE

Combining Theorem 5.1 and existence Theorem 2.6, we get

THEOREM 5.2. *The problem*

$$-\Delta u + \lambda |u|^{p-2} u = h(\mathbf{x}) \qquad \text{in } \mathbb{R}^N,$$

has a unique generalized solution $u \in \mathcal{E}_p$ *for any* N, $1 < p < +\infty$, $\lambda > 0$, *and* $h \in L_{p/(p-1)}(\mathbb{R}^N)$.

5.3 WEAKLY LOWER SEMICONTINUOUSNESS

For the sake of completeness we present here the result that discloses the role of monotonicity for the study of existence problems.

As we have seen, for such problems the property of weakly lower semicontinuousness for a corresponding functional is often crucial.

THEOREM 5.3. *Let* X *be a Banach space, and a functional* $J \in C^1(X)$ *be monotonic, i.e.,*

$$\langle J'(u_1) - J'(u_2), u_1 - u_2 \rangle \geq 0$$

for any $u_1, u_2 \in X$.

Then J *is weakly lower semicontinuous.*

Proof. The proof is a generalization of that for Lemma 2.4. See also Vainberg [2].

Chapter 2
Variational Methods for Eigenvalue Problems

Up to now we have studied problems of a coercive type. Investigation of noncoercive problems requires other methods. One of the ways is the reduction of an original elliptic problem to a new one with a free parameter (eigenvalue) and the investigation of this new problem, for example, by the method of a conditional extremum.

Of course, not every problem with a parameter can be reduced to a problem without a parameter. Only an equation with some homogeneity possesses this characteristic. However, equations with a free parameter themselves are of separate interest. This chapter is devoted to such equations.

§6. Abstract Theorems

In this section we set up the problem and state a number of abstract theorems permitting us to study the solvability of elliptic problems with a parameter by the method of a conditional minimum. Applications of these abstract theorems are given in the following paragraphs.

6.1 Setting up the Problem. Basis of the Method

Here and later on in this section, we consider the problem with a parameter λ of the following form:

$$-\Delta u + f_1(\mathbf{x}, u) + \lambda f_2(\mathbf{x}, u) = 0 \qquad \text{in } \mathbb{R}^N, \tag{6.1}$$

where f_i are Carathéodory functions with $f_1(\mathbf{x}, 0) \equiv f_2(\mathbf{x}, 0) \equiv 0$.

Problem (6.1), for all λ, possesses an obvious trivial solution $u \equiv 0$. We shall be interested in such λ for which problem (6.1) has a nontrivial generalized solution.

In what follows we shall omit the name generalized understanding always that the solutions considered may indeed not be classical and that the boundary condition of (6.1) is also understood in the generalized form.

The value of λ for which problem (6.1) has a nontrivial solution is called an eigenvalue, and the nontrivial solution corresponding to it is called an eigenfunction of (6.1).

In the study of problem (6.1), we shall use the conditional extremum method, which is based on the following classical theorem on a conditional critical point.

THE LAGRANGE THEOREM. *Let X be a Banach space, and J_1, $J_2 : X \to \mathbb{R}$ be functionals of the class $C^1(X)$. Let $u \in X$ be a point of a minimum of the functional J_1 under the condition $J_2 = C_0$, where C_0 is a constant, that is,*

$$J_1(u) = \min_{\substack{v \in X \\ J_2(v)=C_0}} J_1(v).$$

Then there exist constants λ_1 and λ_2 not simultaneously equal to zero such that

$$\lambda_1 J_1'(u) + \lambda_2 J_2'(u) = 0 \qquad \text{in } X^*. \tag{6.2}$$

According to the conditions of the theorem, at least one of the parameters λ_1 or λ_2 is not equal to zero, therefore, identity (6.2) may be rewritten in one of the following two forms:

$$J_1'(u) + \lambda J_2'(u) = 0, \tag{6.3}$$

$$J_2'(u) + \lambda J_1'(u) = 0, \tag{6.4}$$

where λ is a constant. It is obvious that if

$$J_2'(u) \neq 0, \tag{6.5}$$

then (6.2) implies $\lambda_1 \neq 0$ and, consequently, (6.2) and (6.3) become equivalent. Analogously, if

$$J_1(u) \neq 0, \tag{6.6}$$

then $\lambda_2 \neq 0$, and (6.2) and (6.4) become equivalent in this case.

Instead of (6.5) we establish a simpler sufficient condition that guarantees (6.3).

LEMMA 6.1. *Let for all $u \in X$ with $J_2(u) = C_0$,*

$$\langle J_2'(u), u \rangle \neq 0.$$

Then $\lambda_1 \neq 0$, and (6.2) implies (6.3).

The sufficient condition for (6.4) may be formulated analogously.

Sometimes the knowledge of the sign of $\lambda_1\lambda_2$ is helpful.

LEMMA 6.2. *Let for any $u \in X$ with $J_2(u) = C_0$,*

$$\langle J_1'(u), u \rangle \langle J_2'(u), u \rangle > 0. \tag{6.7}$$

Then $\lambda_1\lambda_2 < 0$.

Proof. From (6.2), for any point $u_0 \in X$ of a minimum,

$$\lambda_1 \langle J_1'(u_0), u_0 \rangle + \lambda_2 \langle J_2'(u_0), u_0 \rangle = 0. \tag{6.8}$$

Lemma 6.1 guarantees that $\lambda_1 \neq 0$; $\lambda_2 \neq 0$ because (6.7) gives

$$\langle J_1'(u_0), u_0 \rangle \neq 0;$$

$\lambda_1\lambda_2 > 0$ is impossible because of (6.7) and (6.8). □

6.2 A Theorem on a Conditional Minimum: The First Variant

Let us start formulating and proving theorems providing the existence of such λ_1, $\lambda_2 \in \mathbb{R}$ and $u \in X$ for which (6.2) is true. For these purposes we introduce some definitions.

Definition. A functional $J_1 : X \to \mathbb{R}$ is called bounded from below on X under the condition

$$J_2(u) = C_0 \tag{6.9}$$

if there exists a constant M such that

$$J_1(u) \geq M$$

holds for all $u \in X$ satisfying (6.9).

Definition. A functional $J_1 : X \to \mathbb{R}$ is called weakly lower semicontinuous on X under condition (6.9) if one can choose from any weakly convergent sequence $\{u_n\} \subset X$,

$$u_n \rightharpoonup u \qquad \text{in } X,$$

with $J_2(u_n) = C_0$, a subsequence $\{u_n\}$ such that

$$\lim_{n\to\infty} J_2(u_n) \geq J_2(u).$$

Definition. A functional $J_1 : X \to \mathbb{R}$ is called coercive on X under condition (6.9) if for any constant M there exists a constant $C(M)$ such that, for all $u \in X$, the relations

$$|J_1(u)| \leq M,$$
$$J_2(u) = C_0,$$

(if such u exist) imply

$$\|u\|_X \leq C(M).$$

Definition. A functional $J_2 : X \to \mathbb{R}$ is called weakly continuous on X if from any weakly convergent sequence $\{u_n\} \subset X$, one can choose a subsequence $\{u_n\}$ such that

$$\lim_{n\to\infty} J_2(u_n) = J_2(u).$$

The following is the theorem on a conditional minimum of a functional.

THEOREM 6.3. *Let X be a reflexive Banach space, and $J_1, J_2 : X \to \mathbb{R}$ be some functionals.*

Let the set

$$\{u \in X, J_2(u) = C_0\} \tag{6.10}$$

be not empty.

Let the functional J_1 be

(i) bounded from below,
(ii) weakly lower semicontinuous,
(iii) coercive

on X under condition (6.9).

(I) Let the functional J_2 be weakly continuous on X.

Then the problem

$$\{u \in X, \min J_1(u), J_2(u) = C_0\} \tag{6.11}$$

has a solution.

Proof. In accordance with (i) and the definition of a conditional infimum there exists a conditionally minimizing sequence $\{u_n\} \subset X$, i.e., such a sequence that

$$\lim_{n\to\infty} J_1(u_n) = I, \tag{6.12}$$

$$J_2(u_n) = C_0, \tag{6.13}$$

where

$$I = \inf_{\substack{u\in X \\ J_2(u)=C_0}} J_1(u). \tag{6.14}$$

From (6.12) there exists a constant $\mathbf{C}$ such that

$$|J_1(u_n)| \leq \mathbf{C} \tag{6.15}$$

holds for all u_n. Because of the conditional coerciveness of J_1 and (6.13), (6.15),

$$\|u_n\|_X \leq \mathbf{C}.$$

Because X is reflexive, we can choose a subsequence with

$$u_n \rightharpoonup u \quad \text{in } X \qquad \text{as } n \to \infty. \tag{6.16}$$

Owing to the weak lower semicontinuousness of J_1, we have up to a subsequence

$$I = \lim_{n\to\infty} J_1(u_n) \geq J_1(u). \tag{6.17}$$

Owing to the weak continuousness of J_2,

$$\lim_{n\to\infty} J_2(u_n) = J_2(u). \tag{6.18}$$

Since, according to the definition,

$$J_1(u) \geq I$$

for all $u \in X$ with $J_2(u) = C_0$, we have, taking into consideration (6.13), (6.17), and (6.18),

$$J_1(u) = I, \qquad J_2(u) = C_0,$$

that is, the conditional minimum I of J_1 is assumed at the point $u \in X$. □

From Lemma 6.2 and the Lagrange theorem, the following can be easily derived.

THEOREM 6.4. *Let the conditions of Theorem 6.3 be satisfied. Let functionals J_1, J_2 belong to $C^1(X)$. Let condition (6.7) be satisfied as in Lemma 6.2.*

Then there exists $\lambda < 0$ such that the equation

$$J_1'(u) + \lambda J_2'(u) = 0 \qquad \text{in } X^* \tag{6.19}$$

has a solution $u \in X$ with $J_2(u) = C_0$.

6.3 COMMENTS

Theorem 6.2 is classical and it works well in bounded domains. However, even a very good functional often fails being weakly continuous on $\mathbb{R}^N$, i.e., does not satisfy condition (I).

As an example, consider the functional $J_2 : H_0^1(\Omega) \to \mathbb{R}$,

$$J_2(u) = \|u\|_p^p, \tag{6.20}$$

where $N > 2$, $1 < p < 2^*$. If Ω is bounded, condition (I) is satisfied because

$$u_n \rightharpoonup u \qquad \text{in } H_0^1(\Omega)$$

yields by virtue of the Kondrashov theorem

$$u_n \to u \qquad \text{in } L_p(\Omega), \tag{6.21}$$

and the continuousness of functional (6.20) together with (6.21) gives $J_2(u_n) \to J_2(u)$.

If $\Omega = \mathbb{R}^N$, this is not true. Indeed, consider a sequence $\{u_n\} \subset H_0^1(\mathbb{R}^N)$ of the form

$$u_n = u_0(\mathbf{x} + \mathbf{e}n),$$

where u_0 is a fixed function from $\mathcal{D}(\mathbb{R}^N)$ and $\mathbf{e}$ is a basis vector in $\mathbb{R}^N$. It is easy to see that for all subsequences,

$$u_n \not\to 0 \qquad \text{in } L_p(\mathbb{R}^N)$$

in spite of

$$\begin{aligned} u_n &\rightharpoonup 0 && \text{in } H_0^1(\mathbb{R}^N), \\ u_n &\to 0 && \text{almost everywhere in } \mathbb{R}^N, \\ u_n &\to 0 && \text{in } L_p(B_R) \qquad \text{for all R.} \end{aligned}$$

This means that J_2 does not satisfy condition (I) on $H_0^1(\mathbb{R}^N)$.

In order to get over these difficulties one has to weaken this condition.

6.4 A Theorem on a Conditional Minimum: The Second Variant

THEOREM 6.5. *Let X be a reflexive Banach space, and $J_1, J_2 : X \to \mathbb{R}$ be functionals.*

Let set (6.10) be not empty. Let conditions (i), (ii), (iii) be satisfied for J_1.

(II) Let the functional J_2 be weakly lower semicontinuous on X,

(III) Let, for any $u \in X$ with $J_2(u) < C_0$, there exist a $v \in X$ such that

$$J_1(v) < J_1(u), \qquad J_2(v) = C_0.$$

Then problem (6.11) has a solution.

Proof. Since conditions (i), (ii), (iii) are satisfied, the existence of a conditionally minimizing sequence, with I as in (6.14), for which (6.16) and (6.17) hold, may be proved as in the previous section.

Because condition (I) is replaced by (II) and (III), the rest of the proof is altered. We have from (II)

$$C_0 = \lim_{n\to\infty} J_2(u_n) \geq J_2(u) \tag{6.22}$$

for a subsequence. Now, if $J_2(u) < C_0$, we find, making use of (III), a $v \in X$ such that

$$\begin{aligned} &J_1(v) < J_1(u), \\ &J_2(v) = C_0. \end{aligned} \tag{6.23}$$

Then from (6.17) and (6.23),

$$\begin{aligned} &J_1(v) < I = \inf_{\substack{u\in X \\ J_2(u)=C_0}} J_1(u), \\ &J_2(v) = C_0, \end{aligned}$$

which contradicts the definition of I. Hence, only the equality is possible in (6.22):

$$J_2(u) = C_0. \tag{6.24}$$

From (6.24) and the definition of I, we have

$$J_1(u) \geq I. \tag{6.25}$$

Comparing (6.17) and (6.25), we get

$$J_1(u) = I. \tag{6.26}$$

Relations (6.24) and (6.26) do constitute the statement of the theorem. □

In accordance with the Lagrange theorem the following is true.

THEOREM 6.6. *Let functionals J_1, J_2 belong to $C^1(X)$. Let the conditions of Theorem 6.5 be satisfied. Let conditions*

$$J_2(u) = C_0, \qquad J_2'(u) = 0,$$

be incompatible for $i = 1$ and $i = 2$.

Then there exists $\lambda > 0$ such that equation (6.19) has a solution $u \in X$ with $J_2(u) = C_0$.

Proof. From the Lagrange theorem

$$\lambda_1 J_1'(u) + \lambda_2 J_2'(u) = 0.$$

If $\lambda_1 = 0$, then $J_2'(u) = 0$, contradicting $J_2(u) = C_0$. Analogously, $\lambda_2 \neq 0$.

Now we show that the case $\lambda = \lambda_2/\lambda_1 < 0$ is impossible. Let, to the contrary, $\lambda < 0$. Then there exists a function $\varphi \in X$ such that

$$\langle J_1'(u), \varphi \rangle < 0, \qquad \langle J_2'(u), \varphi \rangle < 0.$$

Consequently, there exists a positive constant ε such that

$$J_1(u + \varepsilon\varphi) < J_1(u),$$
$$J_2(u + \varepsilon\varphi) < C_0.$$

Hence, by condition (III), there is a $v \in X$ with

$$J_1(v) < J_1(u) = I, \qquad J_2(v) = C_0,$$

which contradicts the fact that u solves variational problem (6.11). □

6.5 A THEOREM ON A CONDITIONAL MINIMUM: THE THIRD VARIANT

We present here a version of Theorem 6.5 in which condition (II) is replaced by a weaker one.

THEOREM 6.7. *Let X be a reflexive Banach space, and $J_1, J_2 : X \to \mathbb{R}$ be functionals.*

Let set (6.10) be not empty. Let conditions (i), (ii), (iii) for J_1, and (II) for J_2 be satisfied. Assume that

(IV) there exists $\varepsilon > 0$ such that for any $u \in X$ with

$$J_1(u) - I \leq \varepsilon, \qquad (6.27)$$
$$J_2(u) < C_0,$$

there is a $v \in X$ such that

$$J_1(v) < J_1(u), \qquad J_2(v) = C_0.$$

Then problem (6.11) has a solution.

Proof. Because conditions (i), (ii), (iii) are satisfied, the existence of a conditionally minimizing sequence with I as in (6.14), for which (6.16) and (6.17) hold, is proved as in the proof of Theorem 6.3.

From (6.12) we have

$$J_1(u_n) - I \leq \varepsilon \tag{6.28}$$

starting from some n, where we may choose ε according to (IV). Relations (6.17) and (6.28), give for a subsequence

$$J_1(u) - I \leq \lim_{n\to\infty} J_1(u_n) - I \leq \varepsilon,$$

that is, u satisfies (6.27).

Now, since condition (IV) differs from condition (III) only by the requirement of (6.27) which has just been proved, the rest of the proof repeats the proof of Theorem 6.5 as it proceeds from (6.22). □

6.6 A Theorem on a Conditional Minimum: The Fourth Variant

We present here the theorem with the weakest conditions.

THEOREM 6.8. *Let X be a reflexive Banach space, and $J_1, J_2 : X \to \mathbb{R}$ be some functionals.*

Let set (6.10) be nonempty. Let conditions (i), (ii), (iii), and (II) be satisfied. Assume that

(V) for all $\{u_n\} \subset X$ satisfying

$$\begin{aligned} u_n &\rightharpoonup u \quad \text{in } X,\\ \lim_{n\to\infty} J_1(u_n) &= I,\\ J_2(u_n) &= C_0, \end{aligned}$$

the equality

$$J_2(u) = C_0$$

holds.

Then problem (6.11) has a solution.

Proof. The proof is analogous to that of Theorem 6.3, where (6.18) was a consequence of (6.16) and (I). Here (6.18) is a direct consequence of (6.16), (6.12), (6.13), and (V). □

Remark. Theorem 6.8 is rather easy to prove; it is little more than a simple reformulation of the definition. But verifying its conditions in concrete examples is often more difficult.

In accordance with Lemma 6.2 and the Lagrange theorem the following is true.

THEOREM 6.9. *Let the conditions of Theorem 6.8 be satisfied. Let functionals J_1, J_2 belong to $C^1(X)$. Let condition (6.7) be satisfied as in Lemma 6.2.*

Then there exists $\lambda < 0$ such that equation (6.19) has a solution $u \in X$ with $J_2(u) = C_0$.

Now we consider applications of the abstract theorems of this paragraph to elliptic problems on $\mathbb{R}^N$.

§7. The Equation $-\Delta u + a(\mathbf{x})|u|^{p-2}u - \lambda b(\mathbf{x})|u|^{q-2}u = 0$

In this section we study the indicated equation by the methods of §6. We suppose that λ is positive and a, b are nonnegative.

7.1 SETTING UP THE PROBLEM. WEIGHTED SPACES

Consider the following problem:

$$-\Delta u + a(\mathbf{x})|u|^{p-2}u - \lambda b(\mathbf{x})|u|^{q-2}u = 0 \qquad \text{in } \mathbb{R}^N, \tag{7.1}$$

where $1 < p < +\infty$, $1 < q < +\infty$, $a, b : \mathbb{R}^N \to \mathbb{R}$ are nontrivial nonnegative measurable functions.

Spaces more general than the spaces $\mathcal{E}_{p,q}$ are pertinent for the study of such problems. Consider the functional

$$\|u\|_{\mathcal{E}_{p(a),q(b)}} = \begin{cases} \|\nabla u\|_2 + \|a^{1/p}u\|_p + \|b^{1/q}u\|_q & \text{if } N \le 2, \\ \|\nabla u\|_2 + \|u\|_{2^*} + \|a^{1/p}u\|_p + \|b^{1/q}u\|_q & \text{if } N > 2, \end{cases}$$

defined on $\mathcal{D}(\mathbb{R}^N)$. Closure of $\mathcal{D}(\mathbb{R}^N)$ in this norm creates a Banach space that we denote by $\mathcal{E}_{p(a),q(b)}$. Such spaces are called *weighted*, and functions a and b are called weights. The spaces $\mathcal{E}_{p(a),q(b)}$ are reflexive.

Define functionals $J_1, J_2 : \mathcal{E}_{p(a),q(b)} \to \mathbb{R}$,

$$J_1(u) = \frac{1}{2}\|\nabla u\|_2^2 + \frac{1}{p}\|a^{1/p}u\|_p^p,$$
$$J_2(u) = \|b^{1/q}u\|_q^q.$$

Obviously, $J_1, J_2 \in C^1(\mathcal{E}_{p(a),q(b)})$ and

$$\langle J_1(u), \varphi\rangle = \int_{\mathbb{R}^N} (\nabla u, \nabla\varphi)\, d\mathbf{x} + \int_{\mathbb{R}^N} a(\mathbf{x})|u|^{p-2}u\,\varphi\, d\mathbf{x},$$
$$\langle J_2(u), \varphi\rangle = q\int_{\mathbb{R}^N} b(\mathbf{x})|u|^{q-2}u\,\varphi\, d\mathbf{x}$$

for all $\varphi \in \mathcal{E}_{p,(a)q(b)}$.

Now, having defined the functional spaces and functionals, we want to know under which conditions on functions a, b, Theorems 6.3 and 6.4 can be applied.

7.2 NONEMPTINESS OF $\{J_2(u) = 1\}$

We want to verify nonemptiness of the set (6.10). Let for convenience $C_0 = 1$.

LEMMA 7.1. *There exists* $u \in \mathcal{E}_{p(a),q(b)}$ *such that* $J_2(u) = 1$.

Proof. Because $b \geq 0$, $b \not\equiv 0$, there exists $v \in \mathcal{E}_{p(a),q(b)}$ such that $J_2(v) > 0$. Let

$$u = (J_2(v))^{-1/q} v.$$

It is easy to see that

$$J_2(u) = J_2(v)(J_2(v)^{-1/q})^q = 1.$$

□

7.3 PROPERTIES OF THE FUNCTIONAL J_1

We verify conditions (i), (ii), (iii).

LEMMA 7.2. *The functional* J_1 *is bounded from below on* $\mathcal{E}_{p(a),q(b)}$.

Proof. It is obvious because $J_1(u) \geq 0$. □

LEMMA 7.3. *The functional* J_1 *is coercive on* $\mathcal{E}_{p(a),q(b)}$ *under the condition* $J_2 = 1$.

Proof. Choose some constant M in such a way that the set

$$E \equiv \{u \in \mathcal{E}_{p(a),q(b)},\ |J_1(u)| \leq M,\ J_2(u) = 1\}$$

is not empty. We have

$$\|u\|_{\mathcal{E}_{p(a),q(b)}} \leq (2\, J_1(u))^{1/2} + (p\, J_1(u))^{1/p} + J_2(u)^{1/q} \leq (2M)^{1/2} + (pM)^{1/p} + 1$$

for all $u \in E$, which proves the lemma. □

LEMMA 7.4. *The functional* J_1 *is weakly lower semicontinuous on* $\mathcal{E}_{p(a),q(b)}$.

Proof. As in Section 2.3, we may infer that the functional

$$T(u) = \frac{1}{2}\|\nabla u\|_2^2$$

is weakly lower semicontinuous on $\mathcal{E}_{p(a),q(b)}$.

Now let

$$u_n \rightharpoonup u \qquad \text{in } \mathcal{E}_{p(a),q(b)}.$$

Then

$$\|u\|_{\mathcal{E}_{p(a),q(b)}} \leq \mathbf{C}. \tag{7.2}$$

Analogously to (2.18) we may choose a subsequence in such a way that

$$u_n \to u \qquad \text{almost everywhere in } \mathbb{R}^N. \tag{7.3}$$

We see that

$$a\,|u_n|^p \geq 0. \tag{7.4}$$

Conditions (7.2), (7.3) and (7.4) bring us to the assumptions of the Lebesgue-Fatou theorem, which guarantees the weak lower semicontinuousness of the functional

$$\|a^{1/p}u\|_p^p.$$

The functional J_2 is weakly lower semicontinuous as a sum of two such functionals. □

7.4 CONDITIONS ON J_1', J_2'

We verify condition (6.7).

LEMMA 7.5. *For all* $u \in \mathcal{E}_{p(a),q(b)}$ *with* $J_2(u) = 1$,

$$\langle J_1'(u), u\rangle\langle J_2'(u), u\rangle > 0. \tag{7.5}$$

Proof. Obviously, $J_2(u) = 1$ yields

$$\langle J_2'(u), u\rangle = qJ_2(u) = q > 0, \tag{7.6}$$

consequently, in particular, $u \not\equiv 0$; taking into consideration that $a \not\equiv 0$ we get

$$\langle J_1'(u), u\rangle = \|\nabla u\|_2^2 + \|a^{1/p}u\|_p^p > 0. \tag{7.7}$$

Combining (7.6) and (7.7) gives (7.5). □

7.5 COMPACT EMBEDDING. EXISTENCE THEOREMS

We have verified all the conditions of Theorem 6.4 except for (I) not imposing additional restrictions on the functions a and b and exponents p, q. Condition (I) will require new restrictions. Before considering possible variants we formulate some conditions in a form common for functional analysis.

DEFINITION. We say that a Banach space X is *compactly embedded* in a Banach space Y if $X \circlearrowleft Y$ and any sequence weakly convergent in X converges in Y strongly. This is denoted by

$$X \circlearrowleft\circlearrowleft Y.$$

Let $b : \mathbb{R}^N \to \mathbb{R}$ be a nonnegative measurable function and $\operatorname{supp} b = \Omega$. Denote the closure of $\mathcal{D}(\Omega)$ in the norm

$$\|b^{1/q}u\|_q(\Omega)$$

by $L_{q(b)}$. Let $\phi : \mathbb{R}^N \to \mathbb{R}$ be a measurable function. If $\|b^{1/q}\phi\|_q(\Omega)$ is finite, we shall identify ϕ and its reduction $\phi \in L_{q(b)}$.

LEMMA 7.6. *Let the assumptions of Section 7.1 hold. Let $\mathcal{E}_{p(a)} \circlearrowright\circlearrowright L_{q(b)}$. Then the functional J_2 is weakly continuous on $\mathcal{E}_{p(a),q(b)}$.*

Proof. It follows from the definitions. □

Using Lemmas 7.1–7.5 we get in accordance with Theorem 6.3

THEOREM 7.7. *Let the assumptions of Section 7.1 hold. Let $\mathcal{E}_{p(a)} \circlearrowright\circlearrowright L_{q(b)}$. Then the variational problem*

$$\{u \in \mathcal{E}_{p(a),q(b)}, \min J_1(u), J_2(u) = 1\} \tag{7.8}$$

has a solution.

Remark. From Theorem 6.4, the following is true:

THEOREM 7.8. *Let $\mathcal{E}_{p(a)} \circlearrowright\circlearrowright L_{q(b)}$. Then there exist $\lambda > 0$ and nontrivial $u \in \mathcal{E}_{p(a),q(b)}$ such that*

$$-\Delta u + a(\mathbf{x})|u|^{p-2}u - \lambda b(\mathbf{x})|u|^{q-2}u = 0 \qquad \text{in } \mathbb{R}^N. \tag{7.9}$$

Generally, if $N \leq 2$, we can not state that u is a generalized solution of the equation with the boundary condition

$$u \to 0 \qquad \text{as } |\mathbf{x}| \to \infty$$

even in a generalized case, but, of course, this is so if $\mathcal{E}_{p(a),q(b)} \circlearrowright \mathcal{E}_{p_0}$ for some $1 \leq p_0 < +\infty$.

Thus, all we want to know are the conditions on a and b for which the corresponding spaces are compactly embedded.

7.6 COMPACT EMBEDDING THEOREMS

In order to avoid cumbersome notation let, in this section, X and Y be the spaces $\mathcal{E}_{p(a),q(b)}$ or $L_{p(a)}$ with some weights and exponents. Denote $X(\Omega)$, $Y(\Omega)$ the spaces of functions $u \in X, Y$ restricted onto Ω.

THEOREM 7.9. *Let*

$$X(B_R) \circlearrowright\circlearrowright Y(B_R) \tag{7.10}$$

for any R, and

$$\lim_{R\to\infty} \sup_{u \in X\setminus\{0\}} \frac{\|u\|_Y(\mathbb{R}^N \setminus B_R)}{\|u\|_X} = 0. \tag{7.11}$$

Then

$$X \circlearrowright\circlearrowright Y.$$

Proof. From (7.10) and (7.11) there exists a positive constant $\mathbf{C}$ such that

$$\|u\|_Y \leq \mathbf{C}\,\|u\|_X. \tag{7.12}$$

Any function $\varphi \in \mathcal{D}(\mathbb{R}^N)$ belongs to X and Y. Because X and Y were defined as closures of $\mathcal{D}(\mathbb{R}^N)$ in corresponding norms, (7.12) means that X is embedded into Y.

Now we are proving the compactness of this embedding. Let

$$u_n \rightharpoonup u \qquad \text{in } X.$$

Hence, in particular,

$$\|u_n\|_X \leq \mathbf{C}. \tag{7.13}$$

For any $\varepsilon > 0$ there exists, from (7.11) and (7.13), a radius R_ε such that

$$\|u_n\|_Y(\mathbb{R}^N \setminus B_{R_\varepsilon}) \leq \varepsilon\, \mathbf{C}^{-1}\, \|u_n\|_X \leq \varepsilon. \tag{7.14}$$

From continuousness of measure, we have

$$\lim_{R\to\infty} \|u\|_Y(\mathbb{R}^N \setminus B_R) = 0,$$

and thus there exists maybe a new R_ε such that (7.14) holds and additionally

$$\|u\|_Y(\mathbb{R}^N \setminus B_{R_\varepsilon}) \leq \varepsilon. \tag{7.15}$$

On the other hand, for these ε and R_ε, we may find according to (7.10) a number n_ε such that for all $n \geq n_\varepsilon$,

$$\|u - u_n\|_Y(B_{R_\varepsilon}) \leq \varepsilon. \tag{7.16}$$

Using (7.14), (7.15), and (7.16) we, given an ε, may find R_ε and then n_ε such that

$$\|u - u_n\|_Y \leq \|u - u_n\|_Y(B_{R_\varepsilon}) + \|u\|_Y(\mathbb{R}^N \setminus B_{R_\varepsilon}) + \|u_n\|_Y(\mathbb{R}^N \setminus B_{R_\varepsilon}) \leq 3\varepsilon.$$

As ε is arbitrary, the theorem follows. □

7.7 EXAMPLES OF COMPACT EMBEDDINGS

Compact embeddings of different functional spaces are a subject of separate study (e.g., Muckenhoupt [1], Opic and Kufner [1]). We present some examples.

The Kondrashov theorem may be reformulated in new terms as follows.

KONDRASHOV EMBEDDING THEOREM. *Let $1 \leq p < +\infty$. Then*

$$\mathcal{E}_p(B_R) \circlearrowleft\circlearrowright L_q(B_R) \tag{7.17}$$

for all $1 \leq q < 2^*$ and R.

Remark. Unfortunately, (7.17) does not remain true if we replace B_R by $\mathbb{R}^N$. This was actually demonstrated in Section 6.3.

Now we turn back to $\mathbb{R}^N$.

THEOREM 7.10. *Let $N > 2$, $1 < p < q < 2^*$, a, b be nonnegative nontrivial measurable functions, b locally bounded, and*

$$\lim_{R\to\infty} \sup_{\mathbf{x}\in\mathbb{R}^N\setminus B_R} (b(\mathbf{x}))^{2^*-p}(a(\mathbf{x}))^{q-2^*} = 0. \tag{7.18}$$

Then

$$\mathcal{E}_{p(a)} \circlearrowright\circlearrowright L_{q(b)}(\mathbb{R}^N).$$

Proof. We shall make use of Theorem 7.9. Because b is locally bounded, we have from the Kondrashov embedding theorem and the agreement of Section 7.5 that

$$\mathcal{E}_{p(a)}(B_R) \circlearrowright \mathcal{E}_p(B_R) \circlearrowright\circlearrowright L_q(B_R) \circlearrowright L_{q(b)}(B_R),$$

and hence (7.10) holds.

On the other hand, by the Hölder inequality and the Sobolev embedding theorem,

$$\begin{aligned}\int_{\mathbb{R}^N\setminus B_R} b\,|u|^q\,d\mathbf{x} &\le \left(\int_{\mathbb{R}^N\setminus B_R} b^{\frac{2^*-p}{2^*-q}}|u|^p\,d\mathbf{x}\right)^{\frac{2^*-q}{2^*-p}} \left(\int_{\mathbb{R}^N\setminus B_R} |u|^{2^*}\,d\mathbf{x}\right)^{\frac{q-p}{2^*-p}}\\ &\le \mathbf{C}(R)\left(\|a^{1/p}u\|_p + \|\nabla u\|_2\right)^q\end{aligned}$$

with $\mathbf{C}(R) \to 0$ as $R \to \infty$ in accordance with (7.18).

Thus, the assumptions of Theorem 7.9 hold, and the proof is completed. □

THEOREM 7.11. *Let $N > 2$, $1 < p < q < 2^*$, a, b be nontrivial nonnegative measurable functions, b be locally bounded, and*

$$\begin{aligned}\lim_{R\to\infty} \sup_{\mathbf{x}\in\mathbb{R}^N\setminus B_R} \frac{b(\mathbf{x})}{a(\mathbf{x})} &= 0,\\ \lim_{R\to\infty} \sup_{\mathbf{x}\in\mathbb{R}^N\setminus B_R} b(\mathbf{x}) &= 0.\end{aligned} \tag{7.19}$$

Then

$$\mathcal{E}_{p(a)} \circlearrowright\circlearrowright L_{q(b)}(\mathbb{R}^N).$$

Proof. We have from the Hölder inequality and the Sobolev embedding theorem

$$\begin{aligned}\int_{\mathbb{R}^N\setminus B_R} b\,|u|^q\,d\mathbf{x} &\le \left(\|b^{1/p}u\|_p^{p\frac{2^*-q}{2^*-p}}\,\|b^{1/2^*}u\|_{2^*}^{2^*\frac{q-p}{2^*-p}}(\mathbb{R}^N\setminus B_R)\right)\\ &\le \mathbf{C}\left(\sup_{\mathbf{x}\in\mathbb{R}^N\setminus B_R}\left(\frac{b(\mathbf{x})}{a(\mathbf{x})}\right)^{q/p}\|a^{1/p}u\|_p^q + \sup_{\mathbf{x}\in\mathbb{R}^N\setminus B_R}(b(\mathbf{x}))^{q/2^*}\|u\|_{2^*}^q\right)\\ &\le \mathbf{C}(R)\left(\|a^{1/p}u\|_p + \|\nabla u\|_2\right)^q\end{aligned} \tag{7.20}$$

with $\mathbf{C}(R) \to 0$ as $R \to \infty$ in accordance with (7.19).

The rest is analogous to the proof of Theorem 7.10. □

THEOREM 7.12. *Let $N > 2$, $1 < p < +\infty$, $1 < q < 2^*$, and $b \in L_r(\mathbb{R}^N)$ be a nontrivial nonnegative function with*

$$\frac{2^*}{2^* - q} < r \begin{cases} < +\infty, & p \leq q, \\ \leq \frac{p}{p-q}, & p > q. \end{cases}$$

Then

$$\mathcal{E}_p \circlearrowleft\circlearrowright L_{q(b)}(\mathbb{R}^N).$$

Proof. We may find $\varepsilon > 0$ such that for any R

$$\int\limits_{B_R} b\,|u|^q\,d\mathbf{x} \leq \|b\|_r(B_R)\,\|u\|^q_{\frac{rq}{r-1}} \leq \mathbf{C}(B_R)\,\left(\|u\|_p + \|u\|_{2^*-\varepsilon}\right)^q$$

because $p \leq \frac{rq}{r-1} < 2^*$. This in accordance with the Kondrashov embedding theorem provides (7.10).

On the other hand, analogous reasoning gives

$$\int\limits_{\mathbb{R}^N\setminus B_R} b\,|u|^q\,d\mathbf{x} \leq \|b\|_r(\mathbb{R}^N \setminus B_R)\,\|u\|^q_{\frac{rq}{r-1}} \leq \mathbf{C}(R)\,\left(\|u\|_p + \|u\|_{2^*}\right)^q$$

with $\mathbf{C}(R) \to 0$ as $R \to \infty$, which provides (7.11) and completes the proof. □

As a corollary of Theorems 7.10–7.12, and Theorems 7.7 and 7.8 we have

THEOREM 7.13. *Let a and b satisfy the conditions of one of Theorems 7.10–7.12. Then variational problem (7.8) has a solution, and there exist $\lambda > 0$ and a nontrivial $u \in \mathcal{E}_{p(a),q(b)}$ which solve problem (7.9).*

7.8 THE SIGN OF A SOLUTION

It is often necessary to know in applications about the existence of a solution that does not change sign. The method of §6 gives us certain information on this matter. Consider the equation (7.9).

THEOREM 7.14. *Let the conditions of Theorem 7.13 be true. Then there exist $\lambda > 0$ and a nontrivial nonnegative (and nonpositive) $u \in \mathcal{E}_{p(a),q(b)}$ that satisfy (7.9).*

We prove the existence of a nonnegative solution.

Proof. Theorem 7.8 was a consequence of Theorem 7.7, which dealt with variational problem (7.8). We claim the following

LEMMA 7.15. *Let u be a solution of (7.8). Then function $|u|$ is also a solution of (7.9).*

Proof. If $u \in \mathcal{E}_{p(a),q(b)}$ then $|u| \in \mathcal{E}_{p(a),q(b)}$ too (see Gilbarg and Trudinger [1]). Because

$$J_1(u) = J_1(|u|), \qquad J_2(u) = J_2(|u|),$$

we get the statement. □

Now because the existence of a solution u to (7.8) was proved and $|u| \geq 0$, Lemma 7.15 guarantees the existence of a nontrivial nonnegative solution to (7.8) and, consequently, to (7.9). □

Remark. Under certain conditions on a, b we may assert that any nonnegative solution of (7.9) is positive. This fact is a consequence of the Harnack inequality (Trudinger [1]) but the details are beyond our scope here.

7.9 A MORE GENERAL THEOREM

The results of the preceding sections may be generalized on problems of the form

$$-\Delta u + f_1(\mathbf{x}, u) + \lambda f_2(\mathbf{x}, u) = 0 \qquad \text{in } \mathbb{R}^N, \tag{7.21}$$

where f_i are Carathéodory functions with $f_1(\mathbf{x}, 0) \equiv f_2(\mathbf{x}, 0) \equiv 0$.

DEFINITION. We say a Carathéodory function $g : \mathbb{R}^N \times \mathbb{R} \to \mathbb{R}$ belongs to the class $K_{p(a),q(b)}$ if there are nonnegative measurable functions $a, b : \mathbb{R}^N \to \mathbb{R}$, functions $g_1 \in L_{p/(p-1),(a)}(\mathbb{R}^N)$ and $g_2 \in L_{q/(q-1),(b)}(\mathbb{R}^N)$, exponents $1 < p, q < +\infty$, and a constant $\mathbf{C} > 0$ such that

$$|g(\mathbf{x}, t)| \leq \mathbf{C}\left(a(\mathbf{x})\,|t|^{p-1} + b(\mathbf{x})\,|t|^{q-1}\right) + g_1(\mathbf{x}) + g_2(\mathbf{x})$$

for almost all $\mathbf{x} \in \mathbb{R}^N$ and all $t \in \mathbb{R}$. Assume also $K_{p(a)} = K_{p(a),p(a)}$.

Denote

$$F_i(\mathbf{x}, t) = \int_0^t f_i(\mathbf{x}, s)\, ds.$$

THEOREM 7.16. *Let $f_1 \in K_{p(a)}$ and a nontrivial $f_2 \in K_{q(b)}$, where $1 < p < q < 2^*$ and a, b satisfy the conditions of one of Theorems 7.10–7.12. Let there be a constant $C > 0$ such that*

$$F_1(\mathbf{x}, t) \geq C\, a(\mathbf{x})\, |t|^p.$$

Then there exists a λ such that problem (7.21) has a nontrivial solution $u \in \mathcal{E}_{p(a),q(b)}$.

Proof. The proof is analogous to the proof of Theorem 7.14. □

§8. Radial Solutions

We saw that the spaces $\mathcal{E}_{p,q}$ are not compactly embedded into $L_r(\mathbb{R}^N)$ with any p, q, and r for which we defined these spaces, which does not allow us to use the corresponding existence theorems. However, if we consider some subspaces of the above-mentioned spaces, the situation may improve. Such a case arises if functions a, b depend only on $|\mathbf{x}|$. In this situation we shall find solutions to appropriate equations in the class of radial functions.

8.1 Radial Lemmas

As we study radial solutions, we need to know some facts about radial functions, collected in the following radial lemmas, which are due to Coleman, Glazer, and Martin [1] and Strauss [1].

Lemma 8.1. *Let $N > 2$, and $u \in \mathcal{E}_{2^*}$ be a radial function. Then*

$$|u(\mathbf{x})| \leq \frac{1}{[(N-2)S_N]^{1/2}} \|\nabla u\|_2 |\mathbf{x}|^{\frac{2-N}{2}}, \tag{8.1}$$

where S_N is the area of a unit sphere in $\mathbb{R}^N$.

Proof. Let $u \in \mathcal{D}(\mathbb{R}^N)$. Then

$$|u(R) - u(r)| \leq \int_r^R \left| \frac{\partial u(\rho)}{\partial \rho} \right| d\rho,$$

where $\rho = |\mathbf{x}|$. Using the Hölder inequality, we get

$$\begin{aligned} |u(R) - u(r)| &\leq \left(\int_r^R \left(\frac{\partial u}{\partial \rho} \right)^2 \rho^{N-1}\, d\rho \right)^{1/2} \left(\int_r^R \rho^{1-N}\, d\rho \right)^{1/2} \\ &\leq \left| \frac{R^{2-N} - r^{2-N}}{(N-2)S_N} \right|^{1/2} \|\nabla u\|_2. \end{aligned}$$

Passing to the limit $R \to +\infty$ and taking into consideration that the support of $u \in \mathcal{D}(\mathbb{R}^N)$ is bounded, we get (8.1).

Now if $\{u_n\} \subset \mathcal{D}(\mathbb{R}^N)$ is a sequence strongly convergent to u in $\mathcal{E}_{2^*}$, then from (8.1)

$$|u_n(r) - u_m(r)| \leq \mathbf{C} r^{\frac{2-N}{2}} \|\nabla(u_n - u_m)\|_2 \to 0 \qquad \text{as } n \geq m \to \infty.$$

Hence (8.1) remains true for all $u \in \mathcal{E}_{2^*}$. □

LEMMA 8.2. *Let $N > 1$, and $u \in H_0^1(\mathbb{R}^N)$ be radial. Then*

$$|u(\mathbf{x})| \le S_N^{-1/2}(\|\nabla u\|_2 + \|u\|_2)\, |\mathbf{x}|^{\frac{1-N}{2}}. \tag{8.2}$$

Proof. It is sufficient, analogously to Lemma 8.1, to consider only $u \in \mathcal{D}(\mathbb{R}^N)$. For any such u

$$|(u^2)'| = 2|u'u| \le u^2 + (u')^2.$$

Consequently,

$$\begin{aligned} |u^2(R) - u^2(r)| &\le \int_r^R |(u^2)'|\, d\rho \le r^{-N+1} \int_r^R (u^2 + (u')^2)\rho^{N-1}\, d\rho \\ &= S_N^{-1}\left(\|\nabla u\|_2^2 + \|u\|_2^2\right) r^{-N+1}. \end{aligned} \tag{8.3}$$

Passing to the limit $R \to \infty$ on the right-hand side of (8.3), we get (8.2). □

8.2 EXISTENCE RESULTS

As an example consider the problem

$$-\Delta u + a(|\mathbf{x}|)|u|^{p-2}u - \lambda|u|^{q-2}u = 0 \qquad \text{in } \mathbb{R}^N, \tag{8.4}$$

where $a : \mathbb{R}^+ \to \mathbb{R}$ is a measurable nonnegative locally bounded function with

$$\lim_{R\to\infty} \inf_{\mathbf{x}\in\mathbb{R}^N\setminus B_R} a(|\mathbf{x}|) > 0. \tag{8.5}$$

THEOREM 8.3. *Let $N > 2$, $1 < p < q < 2^*$, and (8.5) be satisfied. Then there exists $\lambda > 0$ such that (8.4) has a nontrivial radial solution $u \in \mathcal{E}_{p(a),q}$.*

THEOREM 8.4. *Let $N = 2$, $2 = p < q < +\infty$, and (8.5) be satisfied. Then there exists $\lambda > 0$ such that (8.4) has a nontrivial radial solution $u \in \mathcal{E}_{p(a),q}$.*

Proof. Let $\mathcal{E}^{\text{rad}}_{p(a),q}$ be the subspace of $\mathcal{E}_{p(a),q}$ consisting of radial functions. Define functionals $J_1, J_2 : \mathcal{E}^{\text{rad}}_{p(a),q} \to \mathbb{R}$,

$$J_1(u) = \frac{1}{2}\|\nabla u\|_2^2 + \frac{1}{p}\|a^{1/p}u\|_p^p,$$
$$J_2(u) = \|u\|_q^q.$$

Because $\mathcal{E}^{\text{rad}}_{p(a),q}$ is an invariant space to equation (8.4) (see Appendix C), any critical point of the functional $J_1 + \lambda J_2 : \mathcal{E}^{\text{rad}}_{p(a),q} \to \mathbb{R}$ is a solution to (8.4). This means that if the conditions of Theorem 6.4 are satisfied, Theorem 8.3 or 8.4 is naturally proved. Because $\mathcal{E}^{\text{rad}}_{p(a),q}$ is a subspace of $\mathcal{E}_{p(a),q}$, conditions (i), (iii), (iii) of Section 6.2 are valid because they were verified for the functionals defined on $\mathcal{E}_{p(a),q}$ (see Lemmas 7.2–7.4); the nonemptiness of (6.10) may be proved as was done earlier (see Lemma 7.1). As in the preceding paragraph, condition (I) of Section 6.2 will be a consequence of Lemma 7.6 and compact embeddings of corresponding spaces. From conditions on a, $\mathcal{E}^{\text{rad}}_{p(a)} \circlearrowleft \mathcal{E}^{\text{rad}}_p$, so all we need now is to show that the following is true.

LEMMA 8.5. *Under conditions of Theorem 8.3 or 8.4,*

$$\mathcal{E}_p^{\mathrm{rad}} \circlearrowleft\circlearrowleft L_q^{\mathrm{rad}}.$$

Proof. We consider the conditions of Theorem 8.3.

Embedding $\mathcal{E}_p^{\mathrm{rad}}(B_R) \circlearrowleft\circlearrowleft L_q^{\mathrm{rad}}(B_R)$ follows from the Kondrashov theorem.

On the other hand, due to radial Lemma 8.1, we have for sufficiently large r,

$$|u(r)|^q = |u(r)|^p |u(r)|^{q-p} \le \mathbf{C}\, |u(r)|^p\, \|\nabla u\|_2^{q-p}\, r^{-\frac{N-2}{2}(q-p)}.$$

Then, by the Young inequality,

$$\frac{\|u\|_q^q(\mathbb{R}^N \setminus B_r)}{\|u\|_p^q + \|\nabla u\|_2^q} \le \mathbf{C}\, r^{-\frac{N-2}{2}(q-p)}$$

and tends to zero as $r \to \infty$.

Taking into account Theorem 7.9, we get the desired embedding.

Thus, Lemma 8.5 under the conditions of Theorem 8.3 is proved.

The case of the conditions of Theorem 8.4 may be considered analogously with the use of radial Lemma 8.2 instead of 8.1. □

Lemma 8.5 is proved; so are Theorems 8.3 and 8.4. □

8.3 THE STRAUSS LEMMA

We present here one preliminary result which develops in a sense Lemma 8.5 and will be useful later.

THE STRAUSS LEMMA. *Let $P, Q : \mathbb{R} \to \mathbb{R}$ be continuous functions with $Q(s) \neq 0$ for $s \neq 0$ such that*

$$\lim_{|s|\to 0} \left| \frac{P(s)}{Q(s)} \right| = 0, \qquad \lim_{|s|\to \infty} \left| \frac{P(s)}{Q(s)} \right| = 0.$$

Then, for any sequence $\{u_n\}$ of measurable functions on $\mathbb{R}^N$ satisfying

(1) $u_n \to u$ *almost everywhere in* $\mathbb{R}^N$ *as* $n \to \infty$,
(2) $\|Q(u_n)\|_1 \le \mathbf{C}$ *uniformly with respect to* n,
(3) $u_n(\mathbf{x}) \to 0$ *as* $|\mathbf{x}| \to +\infty$ *uniformly with respect to* n,

the following is true:

$$P(u_n) \to P(u) \qquad \text{in } L_1(\mathbb{R}^N) \qquad \text{as } n \to \infty.$$

Proof. See Appendix B. □

§9. The Equation $-\Delta u + \lambda f(u) = 0$

In this section, we consider the problem

$$-\Delta u + \lambda f(u) = 0 \qquad \text{in } \mathbb{R}^N. \tag{9.1}$$

We see that equation (9.1) is invariant with regard to translations in $\mathbb{R}^N$. Indeed, if u is a solution of (9.1) then $u(\,.\, + \mathbf{x}_0)$, where $\mathbf{x}_0 \in \mathbb{R}^N$ is some constant vector, is a solution as well. Equation (9.1) is also invariant with respect to rotations. This implies that we may search for solutions to (9.1) among radial functions — i.e., functions with $u(\mathbf{x}) = u(|\mathbf{x}|)$ — where any point of $\mathbb{R}^N$ may be chosen as the origin.

In this paragraph we follow mainly Berestycki and Lions [1]. See also Brezis and Lieb [1].

9.1 THE FIRST SAMPLE

For the rest of this section we suppose that $N > 2$. Let $f : \mathbb{R} \to \mathbb{R}$ be a continuous function such that

(a_1) $f(0) = 0$,

(a_2) there exists $t_0 > 0$ such that $F(t_0) < 0$ where

$$F(t) = \int_0^t f(s)\, ds.$$

Let also

(b)

$$0 \le \varliminf_{t \to +0} \frac{F(t)}{t^{2^*}},$$

(c)

$$0 \le \varliminf_{t \to +\infty} \frac{F(t)}{t^{2^*}},$$

(d)

$$f \in K_{2^*}.$$

THEOREM 9.1. *Let conditions (a), (b), (c), (d) hold. Then there exist $\lambda > 0$ and a nontrivial nonnegative $u \in \mathcal{E}_{2^*}$ that solve (9.1).*

Proof. Let $X = \mathcal{E}_{2^*}^{\mathrm{rad}}$, $C_0 = -1$. Define functionals $J_1, J_2 : \mathcal{E}_{2^*}^{\mathrm{rad}} \to \mathbb{R}$ in the form

$$J_1(u) = \|\nabla u\|_2^2,$$

$$J_2(u) = \int_{\mathbb{R}^N} F(u)\, d\mathbf{x} \tag{9.2}$$

and make use of Theorem 6.6. All we need is to verify its assumptions. In full analogy with Section 7.8 we shall search for a nonnegative solution; thus suppose without losing generality that $F(t) \equiv F(-t)$.

Preliminary, we prove one simple result which often will be referred to.

9.2 SCALING

Let $u \in \mathcal{E}_{2^*}$. For any $\sigma > 0$ define $u_\sigma(\mathbf{x}) = u(\mathbf{x}/\sigma)$.

LEMMA 9.2.

$$J_1(u_\sigma) = \sigma^{N-2} J_1(u),$$
$$J_2(u_\sigma) = \sigma^N J_2(u).$$

Proof. This lemma follows from obvious identities:

$$J_1(u_\sigma) = \int_{\mathbb{R}^N} |\nabla_{\mathbf{x}} u(\mathbf{x}/\sigma)|^2 \, d\mathbf{x} = \sigma^{N-2} \int_{\mathbb{R}^N} |\nabla_{\mathbf{x}/\sigma} u(\mathbf{x}/\sigma)|^2 \, d(\mathbf{x}/\sigma) = \sigma^{N-2} J_1(u),$$

$$J_2(u_\sigma) = \int_{\mathbb{R}^N} |u(\mathbf{x}/\sigma)|^2 \, d\mathbf{x} = \sigma^N \int_{\mathbb{R}^N} |u(\mathbf{x}/\sigma)|^2 \, d(\mathbf{x}/\sigma) = \sigma^N J_2(u). \qquad \square$$

We now verify the assumptions of Theorem 6.6.

9.3 NONEMPTINESS OF $\{J_2 = -1\}$

Choose $C_0 = -1$.

LEMMA 9.3. *The set*

$$\{u \in \mathcal{E}_{2^*}^{\mathrm{rad}}, J_2(u) = -1\}$$

is not empty.

Proof. According to condition (a_2) there exists $t_0 > 0$ such that $F(t_0) < 0$. Consider the class of functions $w_R : \mathbb{R} \to \mathbb{R}$ of the form

$$w_R(r) = \begin{cases} t_0, & r \le R, \\ t_0(R - r + 1), & R < r \le R + 1, \\ 0, & r \ge R + 1. \end{cases}$$

We have

$$S_N^{-1} J_2(w_R) = \int_0^R F(t_0) r^{N-1} \, dr + \int_R^{R+1} F(t_0(R - r + 1)) r^{N-1} \, dr$$
$$= \frac{1}{N} F(t_0) \, R^N + \frac{1}{N}((R+1)^N - R^N) F(t_1(R)),$$

where $0 \le t_1(R) \le t_0$. Obviously,

$$(R+1)^N - R^N = \sum_{i=0}^{N-1} C_i R^i,$$

where C_i are some constants. Consequently,

$$\frac{S_N^{-1} J_2(w_R)}{R^N} \to \frac{1}{N} F(t_0) < 0$$

as $R \to \infty$. Hence there exists a positive constant ε such that

$$J_2(u) = -\varepsilon.$$

Assuming

$$\sigma = \varepsilon^{-1/N}$$

we get, from Lemma 9.2,

$$J_2(u_\sigma) = -1. \qquad \square$$

9.4 Properties of the Functional J_1

We verify conditions (i), (ii), (iii) of Section 6.2.

LEMMA 9.4. *The functional J_1 is bounded from below on $\mathcal{E}_{2^*}^{\mathrm{rad}}$.*

Proof This is obvious by force of $J_1(u) \ge 0$. $\square$

LEMMA 9.5. *The functional J_1 is weakly lower semicontinuous on $\mathcal{E}_{2^*}^{\mathrm{rad}}$.*

Proof. See the proof of Lemma 2.3. $\square$

LEMMA 9.6. *The functional J_1 is coercive on $\mathcal{E}_{2^*}^{\mathrm{rad}}$ under the condition $|J_2(u)| = 1$.*

Proof. Choose $M > 0$ in such a way that the set

$$E \equiv \{u \in \mathcal{E}_{2^*}^{\mathrm{rad}}, J_2(u) = -1, |J_1(u)| \le M\}$$

is nonempty. Then for all $u \in E$, from the Sobolev embedding theorem,

$$\|u\|_{2^*} + \|\nabla u\|_2 \le \mathbf{C}\, \|\nabla u\|_2 = \mathbf{C}\, (J_1(u))^{1/2} \le \mathbf{C}\, M^{1/2},$$

and the lemma follows. $\square$

9.5 PROPERTIES OF J_2

We verify condition (II) of Theorem 6.5.

LEMMA 9.7. *The functional J_2 is weakly lower semicontinuous on $\mathcal{E}_{2^*}^{\mathrm{rad}}$.*

Proof. Let

$$u_n \rightharpoonup u \qquad \text{in } \mathcal{E}_{2^*}^{\mathrm{rad}}. \tag{9.3}$$

By the Sobolev embedding theorem,

$$\|\nabla u_n\|_2 + \|u_n\|_{2^*} \leq \mathbf{C}\,\|\nabla u_n\|_2 \leq \mathbf{C}. \tag{9.4}$$

(We again recall that the same symbol $\mathbf{C}$ may denote different positive constants.) Besides, one can choose a subsequence from (9.3) such that

$$u_n \to u \qquad \text{almost everywhere in } \mathbb{R}^N. \tag{9.5}$$

Define functions $F_1, F_2 : \mathbb{R} \to \mathbb{R}$ as

$$F_1 = F^-, \qquad F_2 = -F_1 + F, \tag{9.6}$$

where $F^- = \min\{0, F\}$. We claim that

$$\lim_{n\to\infty} \int_{\mathbb{R}^N} F_1(u_n)\, d\mathbf{x} = \int_{\mathbb{R}^N} F_1(u)\, d\mathbf{x}, \tag{9.7}$$

$$\varliminf_{n\to\infty} \int_{\mathbb{R}^N} F_2(u_n)\, d\mathbf{x} \geq \int_{\mathbb{R}^N} F_2(u)\, d\mathbf{x}. \tag{9.8}$$

To prove (9.7) we make use of the Strauss lemma. Indeed, from (b), (c),

$$\begin{aligned} F_1(t)|_{t\to+0} &= o(t^{2^*}), \\ F_1(t)|_{t\to+\infty} &= o(t^{2^*}). \end{aligned} \tag{9.9}$$

Define $Q : \mathbb{R} \to \mathbb{R}$ by $Q(t) = |t|^{2^*}$. By (9.9),

$$\frac{F_1(t)}{Q(t)} \to 0 \qquad \text{as } t \to 0 \quad \text{or} \quad |t| \to \infty. \tag{9.10}$$

By (9.4),

$$\sup_n \int_{\mathbb{R}^N} |Q(u_n)|\, d\mathbf{x} \leq \mathbf{C} \sup_n \|\nabla u_n\|_2^{2^*} \leq \mathbf{C}. \tag{9.11}$$

According to the radial lemma and (9.4), we have

$$u_n \to 0 \qquad \text{as } |\mathbf{x}| \to \infty \qquad \text{uniformly in } n. \tag{9.12}$$

Using (9.5), (9.10), (9.11), and (9.12), we see that the assumptions of the Strauss lemma are satisfied and we get (9.7).

To prove (9.8) we make use of the Lebesgue-Fatou theorem. Indeed, from (*c*), (*d*),

$$F_2(t)|_{t\to+0} = O(t^{2^*}),$$
$$F_2(t)|_{t\to+\infty} = O(t^{2^*}),$$

consequently, there exists a constant C such that

$$|F_2(t)| \leq C\,|t|^{2^*}$$

holds for all $t \in \mathbb{R}$. Hence, by (9.4),

$$\int_{\mathbb{R}^N} F_2(u_n)\,d\mathbf{x} \leq \mathbf{C}\,\|u_n\|_{2^*}^{2^*} \leq \mathbf{C}. \tag{9.13}$$

Using (9.5) and (9.13) and taking into consideration that $F_2(t) \geq 0$, we can use the Lebesgue-Fatou theorem and get (9.8).

Now, because $F = F_1 + F_2$, (9.7) and (9.8) imply

$$\varliminf_{n\to\infty} \int_{\mathbb{R}^N} F(u_n)\,d\mathbf{x} \geq \int_{\mathbb{R}^N} F(u)\,d\mathbf{x}.$$

□

9.6 CONDITION (*III*)

We verify condition (*III*) of Theorem 6.5 for problem (9.1).

LEMMA 9.8. *For any $u \in \mathcal{E}_{2^*}^{\mathrm{rad}}$, with $J_2(u) < -1$, there exists a $v \in \mathcal{E}_{2^*}$ such that*

$$J_1(v) < J_1(u), \qquad J_2(v) = -1.$$

Proof. Denote

$$\varepsilon = -J_2(u) > 1.$$

Suppose

$$\sigma = \varepsilon^{-1/N} < 1.$$

By Lemma 9.2

$$J_1(u_\sigma) = \varepsilon^{\frac{2-N}{N}} J_1(u) < J_1(u),$$
$$J_2(u_\sigma) = \varepsilon^{-1} J_2(u) = -1.$$

□

9.7 Completing the Proof

From Sections 9.2–9.5 all the conditions of Theorem 6.5 are satisfied, which means that problem (6.11),

$$\left\{u \in \mathcal{E}_{2^*}^{\mathrm{rad}},\ \min J_1(u),\ J_2(u) = -1\right\},$$

has a solution.

The rest of the proof is devoted to showing that conditions

$$J_2(u) = -1, \qquad J_2'(u) = 0,$$

for $i = 1$ and $i = 2$ are incompatible.

Indeed, $J_1'(u) = 0$ means $\nabla u \equiv 0$, that is, $u \equiv \text{Const}$; $J_2'(u) = 0$ means $f(u) \equiv 0$. It is easy to see that the first or the second relation makes $u \in \mathcal{E}_{2^*}^{\mathrm{rad}}$, $J_2(u) = -1$ impossible.

Thus, the conditions of Theorem 6.6 hold, and Theorem 9.1 is proved. □

9.8 An Example for the First Sample

THEOREM 9.9. *Let $N > 2$ and $1 < p < 2^*$. Then there exist $\lambda > 0$ and a nontrivial nonnegative $u \in \mathcal{E}_{2^*}$ satisfying*

$$-\Delta u + \lambda \begin{Bmatrix} |u|^{2^*-2}u, & |u| \le 1, \\ 2-|u|^{p-2}u, & |u| > 1 \end{Bmatrix} = 0 \qquad \text{in } \mathbb{R}^N.$$

9.9 The Second Sample

Now we consider another class of continuous functions f. Suppose that

$$(b') \qquad 0 < \lim_{t \to +0} \frac{F(t)}{t^p} < +\infty \qquad \text{for some exponent } 1 < p < 2^*,$$

$$(d') \qquad f \in K_{p,2^*}.$$

THEOREM 9.10. *Let conditions (a), (b′), (c), (d′) hold. Then there exist $\lambda > 0$ and a nontrivial nonnegative $u \in \mathcal{E}_p$ that solve (9.1).*

Proof. Let $X = \mathcal{E}_p^{\mathrm{rad}}$, $C_0 = -1$. Assume $F(t) \equiv F(-t)$, choose functionals $J_1, J_2 : \mathcal{E}_p^{\mathrm{rad}} \to \mathbb{R}$ in the form of (9.2), and make use of Theorem 6.6 and the reasoning of Section 7.8.

Conditions (i), (ii) may be checked as in the first sample. Verifying (iii) differs.

LEMMA 9.11. *The functional J_1 is coercive on $\mathcal{E}^{\mathrm{rad}}_{p,2^*}$ under the condition $J_2 = -1$.*

Proof. Choose $M > 0$ in such a way that the set

$$E \equiv \{u \in \mathcal{E}^{\mathrm{rad}}_p, J_2(u) = -1, |J_1(u)| \leq M\}$$

is nonempty.

From (b') there exists a positive number m such that

$$m = \lim_{t\to 0} \frac{F(t)}{|t|^p}.$$

Define analogously to (9.7) functions $F_1, F_2 : \mathbb{R} \to \mathbb{R}$ by

$$F_1(t) = (F(t) - m\,|t|^p)^-, \qquad F_2 = -F_1 + F. \tag{9.14}$$

Note that

$$F_2(t) \geq m\,|t|^p. \tag{9.15}$$

Indeed, if t is such that $F(t) \leq m\,|t|^p$, then

$$F_2(t) = -(F(t) - m\,|t|^p)^- + F(t) = -F(t) + m\,|t|^p + F(t) = m\,|t|^p.$$

If $F(t) \geq m\,|t|^p$, then

$$F_2(t) = 0 + F(t) \geq m\,|t|^p.$$

From conditions (b'), (c), (d'), functions F_1, F_2 behave as follows:

$$\begin{aligned} F_1(t)|_{t\to+0} &= o(t^p), \\ F_1(t)|_{t\to+\infty} &= o(t^{2^*}), \\ F_2(t)|_{t\to+0} &\geq mt^p, \\ F_2(t)|_{t\to+\infty} &= O(t^{2^*}). \end{aligned} \tag{9.16}$$

From (9.16), there exists a positive constant C such that, for all $t \in \mathbb{R}$,

$$-F_1(t) \leq \frac{1}{2} F_2(t) + C\,|t|^{2^*}. \tag{9.17}$$

From (9.17) and the Sobolev embedding theorem, the inequality

$$-\int_{\mathbb{R}^N} F_1(u)\,d\mathbf{x} \leq \frac{1}{2}\int_{\mathbb{R}^N} F_2(u)\,d\mathbf{x} + \mathbf{C}\,M^{2^*/2} \tag{9.18}$$

holds for all $u \in E$. Besides, by definition,

$$\int_{\mathbb{R}^N} F_1(u)\,d\mathbf{x} + \int_{\mathbb{R}^N} F_2(u)\,d\mathbf{x} = -1. \tag{9.19}$$

From (9.18) and (9.19), it follows that

$$\int_{\mathbb{R}^N} F_2(u)\,d\mathbf{x} \le \mathbf{C}(M). \tag{9.20}$$

Note that from (9.15),

$$\int_{\mathbb{R}^N} F_2(u)\,d\mathbf{x} \ge m\|u\|_p^p. \tag{9.21}$$

Comparing (9.20) and (9.21) we get for all $u \in E$,

$$\|u\|_p \le \mathbf{C}(M).$$

Finally, we get

$$\|u\|_p + \|\nabla u\|_2 \le \mathbf{C}(M) + (J_1(u))^{1/2} \le \mathbf{C}(M).$$

(We recall that $\mathbf{C}$ may denote different positive constants.) □

9.10 PROPERTIES OF THE FUNCTIONAL J_2

We verify condition (II) of Theorem 6.5.

LEMMA 9.12. *The functional J_2 is weakly lower semicontinuous on $\mathcal{E}_p^{\mathrm{rad}}$.*

Proof. Let

$$u_n \rightharpoonup u \qquad \text{in } \mathcal{E}_p^{\mathrm{rad}}. \tag{9.22}$$

Then

$$\|u_n\|_p + \|\nabla u_n\|_2 \le \mathbf{C}. \tag{9.23}$$

Besides, one can choose a subsequence from (9.22) such that

$$u_n \to u \qquad \text{almost everywhere in } \mathbb{R}^N. \tag{9.24}$$

We claim that

$$\lim_{n\to\infty} \int_{\mathbb{R}^N} F_1(u_n)\,d\mathbf{x} = \int_{\mathbb{R}^N} F_1(u)\,d\mathbf{x}, \tag{9.25}$$

$$\varliminf_{n\to\infty} \int_{\mathbb{R}^N} F_2(u_n)\,d\mathbf{x} \ge \int_{\mathbb{R}^N} F_2(u)\,d\mathbf{x}. \tag{9.26}$$

To prove (9.25) we might make use of the Strauss lemma as in the proof of the preceding theorem. Instead we demonstrate an alternative approach.

By (9.16), for any $\varepsilon > 0$ there is a constant $C_\varepsilon > 0$ and a continuous function $\bar{F}_1 : \mathbb{R} \to \mathbb{R}$ such that for all $t \in \mathbb{R}$,

$$\begin{aligned} F_1(t) &= \bar{F}_1(t) + \varepsilon\,(t^{2^*} + t^p), \\ |\bar{F}_1(t)| &\le C_\varepsilon\,(t^{(2^*+p)/2}). \end{aligned} \tag{9.27}$$

By (9.22), Lemma 8.5 with $q = (2^*+p)/2$, and the Vainberg-Krasnoselskii theorem, for any $\varepsilon > 0$, one can choose n_ε such that for all $n \ge n_\varepsilon$,

$$\int_{\mathbb{R}^N} |\bar{F}_1(u_n) - \bar{F}_1(u)|\,d\mathbf{x} \le \varepsilon. \tag{9.28}$$

From (9.23),

$$\varepsilon \sup_n \int_{\mathbb{R}^N} (|u_n|^{2^*} + |u_n|^p)\,d\mathbf{x} \le \varepsilon\,\mathbf{C} \sup_n(\|\nabla u_n\|_2^{2^*} + \|u_n\|_p^p) \le \varepsilon\,\mathbf{C}. \tag{9.29}$$

Using (9.27), (9.28), and (9.29) we see that for any $\varepsilon > 0$ one can find $n_\varepsilon > 0$ such that the inequality

$$\int_{\mathbb{R}^N} |F_1(u_n) - F_1(u)|\,d\mathbf{x} \le \varepsilon$$

holds for all $n \ge n_\varepsilon$, which proves (9.25).

To prove (9.26) we make use of the Lebesgue-Fatou theorem. Indeed,

$$0 \le F_2(t) \le C(|t|^p + |t|^{2^*}) \tag{9.30}$$

holds for all $t \in \mathbb{R}$. Using (9.23), (9.24), and (9.30) we see that the conditions of the Lebesgue-Fatou theorem and, consequently, (9.26) hold. Summing up (9.25) and (9.26) and taking into consideration (9.14) we complete the proof of Lemma 9.12. □

The rest of the proof of Theorem 9.10 is analogous to that of Theorem 9.1. □

9.11 An Example for the Second Sample

THEOREM 9.13. *Let $N > 2$. Then for any $1 < p < q < 2^*$ there exist $\lambda > 0$ and a nontrivial nonnegative $u \in \mathcal{E}_{p,2^*}$ satisfying*

$$-\Delta u + \lambda(|u|^{p-2}u - |u|^{q-2}u) = 0 \qquad \textit{in } \mathbb{R}^N.$$

9.12 THE THIRD SAMPLE. TRUNCATION

THEOREM 9.14. *Let conditions (a) and (b) be imposed on a locally Lipschitz-continuous function f. Let there be $t_1 > t_0$, where t_0 is from condition (a), such that $f(t_1) = 0$. Let*

$$\lim_{t\to+0} \frac{|f(t)|}{t^{2^*-1}} < C$$

with some $C > 0$. Then there exist $\lambda > 0$ and a nontrivial nonnegative $u \in \mathcal{E}_{2^*}$ satisfying (9.1).

Proof. Define the function

$$\bar{f}(t) = \begin{cases} f(t), & t < t_1, \\ 0, & t \geq t_1. \end{cases} \tag{9.31}$$

Function $\bar{f}$ satisfies the conditions of Theorem 9.1 and, consequently, we shall prove Theorem 9.14 showing that any nonnegative solution $u \in \mathcal{E}_{2^*}^{\mathrm{rad}}$ of equation (9.1), with $\bar{f}$ instead of f, satisfies $|u| < t_1$. By the regularity theory (see Section 1.3) $u \in C^1(\mathbb{R}^N)$, moreover by the radial theory (see Sections 19.1, 19.2) $u \in C^2(\mathbb{R}^N)$. Lemma 8.1 means that u achieves a maximum at some point $r_0 = |\mathbf{x}_0|$, with $u'(r_0) = 0$, $u''(r_0) \leq 0$. If $u''(r_0) = 0$, then from the equation $-\Delta u + \lambda\bar{f}(u) = 0$ and uniqueness theorems for the ODE method of §19 a positive solution $u \in \mathcal{E}_{2^*}$ is the constant function $u = u(r_0)$, which is impossible. Hence, $u''(r_0) < 0$ and, consequently, $\bar{f}(u_{\max}) \neq 0$, that is, $|u| < t_1$. □

§10. The Equation $-\Delta u - \lambda|u|^{p-2}u - b(\mathbf{x})|u|^{q-2}u = 0$

Here we demonstrate an application of Theorem 6.8 on an interesting example.

10.1 SETTING UP THE PROBLEM

Consider the problem

$$-\Delta u - \lambda|u|^{p-2}u - b(\mathbf{x})|u|^{q-2}u = 0 \qquad \text{in } \mathbb{R}^N, \tag{10.1}$$

where $\lambda > 0$ and $b : \mathbb{R}^N \to \mathbb{R}$ is a nonnegative nontrivial measurable function, $b \in L_\infty(\mathbb{R}^N)$. This problem can not be reduced to the preceding ones.

THEOREM 10.1. *Let $N > 2$, $q > 2$,*

$$1 < p < q < \min\{2^*, 2 + \frac{2p}{N}\}, \tag{10.2}$$

and let there be a positive constant C_1 and an exponent α such that

$$b(\mathbf{x}) \geq \frac{C_1}{1+|\mathbf{x}|^\alpha}, \qquad 0 < \alpha < 2 - \frac{N}{p}(q-2)$$

holds for almost all $\mathbf{x} \in \mathbb{R}^N$, *and*

$$\lim_{R\to\infty} \|b\|_\infty(\mathbb{R}^N \setminus B_R) = 0.$$

Then there exist $\lambda > 0$ *and* $u \in \mathcal{E}_{p,q(b)}$ *that solve (10.1).*

The proof of this theorem is given in the following sections.

10.2 THE VARIATIONAL PROBLEM

In accordance with the method of Paragraph 6, associate with (10.1) the variational problem

$$\{u \in \mathcal{E}_{p,q(b)}, \min J_1(u), J_2(u) = 1\} \tag{10.3}$$

where

$$J_1(u) = \frac{1}{2}\|\nabla u\|_2^2 - \frac{1}{q}\|b^{1/q}\|_q^q,$$
$$J_2(u) = \|u\|_p^p.$$

THEOREM 10.2. *Problem (10.3) under the conditions of Theorem 10.1 has a solution.*

We make use of Theorem 6.7 in proving Theorem 10.1. We verify its assumptions.

LEMMA 10.3. *The set*

$$\{u \in \mathcal{E}_{p,q(b)}, J_2(u) = 1\}$$

is not empty.

Proof. The statement is evident (cf. Lemma 7.1). □

LEMMA 10.4. *Functional* J_1 *is weakly lower semicontinuous on* $\mathcal{E}_{p,q(b)}$.

Proof. The functional $(1/2)\,\|\nabla u\|_2^2$ is weakly lower semicontinuous according to Lemma 2.4. The functional $(1/q)\,\|b^{1/q}u\|_q^q$ is weakly semicontinuous on $\mathcal{E}_{p,q(b)}$ in accordance with Lemma 7.6 and Theorem 7.10. Summing up, we get the result. □

LEMMA 10.5. *Functional* J_2 *is weakly lower semicontinuous on* $\mathcal{E}_{p,q(b)}$.

Proof. See the second part of that of Lemma 7.4. □

The other properties are less easy to verify.

10.3 Properties of J_1

Lemma 10.6. *The functional J_1 is bounded from below on $\mathcal{E}_{p,q(b)}$ under the condition*

$$J_2 = 1. \tag{10.4}$$

Proof. From the Hölder inequality we have for all $u \in \mathcal{E}_{p,q(b)}$,

$$\|u\|_q^q = \int\limits_{\mathbb{R}^N} |u|^{p\frac{2^*-q}{2^*-p}} |u|^{\frac{2^*(q-p)}{2^*-p}} \, d\mathbf{x} \le \|u\|_p^{p\frac{2^*-q}{2^*-p}} \|u\|_{2^*}^{\frac{2^*(q-p)}{2^*-p}}. \tag{10.5}$$

According to (10.4), (10.5), and the Sobolev embedding theorem

$$\begin{aligned} J_1(u) &= \frac{1}{2}\|\nabla u\|_2^2 - \frac{1}{q}\|b^{1/q}u\|_q^q \ge \frac{1}{2}\|\nabla u\|_2^2 - \frac{1}{q}\|b\|_\infty \|u\|_q^q \\ &\ge \frac{1}{2}\|\nabla u\|_2^2 - \frac{1}{q}\|b\|_\infty \|u\|_{2^*}^{\frac{2^*(q-p)}{2^*-p}} \ge \frac{1}{2}\|\nabla u\|_2^2 - \mathbf{C}\,\|\nabla u\|_2^{\frac{2^*(q-p)}{2^*-p}}. \end{aligned} \tag{10.6}$$

Because (10.2) is true, $\frac{2^*(q-p)}{2^*-p} < 2$, and the right-hand side of (10.6) is bounded from below, so is the left-hand side. □

Lemma 10.7. *The functional J_1 is coercive on $\mathcal{E}_{p,q(b)}$ under condition (10.4).*

Proof. Let the set

$$E \equiv \{u \in \mathcal{E}_{p,q(b)}, |J_1(u)| \le M, J_2(u) = 1\}$$

be nonempty. Then for all $u \in E$, according to (10.6),

$$M \ge \frac{1}{2}\|\nabla u\|_2^2 - \mathbf{C}\,\|\nabla u\|_2^{\frac{2^*(q-p)}{2^*-p}}. \tag{10.7}$$

Taking into consideration (10.2), we get from (10.7) that

$$\|\nabla u\|_2 \le \mathbf{C}(M). \tag{10.8}$$

Combining (10.4) and (10.8), we get

$$\|\nabla u\|_2 + \|u\|_p = \|\nabla u\|_2 + (J_2(u))^{1/p} \le \mathbf{C}(M).$$

The Sobolev embedding theorem and boundedness of b complete the proof. □

10.4 CONDITION (IV)

Verification of condition (IV) is based on two lemmas which follow.

Define as usual

$$I = \inf_{\substack{u \in \mathcal{E}_{p,q(b)} \\ J_2(u)=1}} J_1(u).$$

LEMMA 10.8. *The estimate*

$$I < 0 \tag{10.9}$$

holds.

Proof. From Lemma 10.7 the set E is not empty. Given u with $J_2(u) = 1$, we may build a family $\{u_\sigma\}_{\sigma>0}$ of functions

$$u_\sigma(\mathbf{x}) = \sigma^{-N/p} u(\mathbf{x}/\sigma).$$

Obviously, as in scaling Lemma 9.2,

$$J_2(u_\sigma) = J_2(u) = 1.$$

Analogously, for $\sigma \geq 1$,

$$J_1(u_\sigma) = \frac{1}{2} \int_{\mathbb{R}^N} |\nabla u_{\mathbf{x}}(\mathbf{x}/\sigma)|^2 \sigma^{-\frac{2N}{p}}\, d\mathbf{x} - \frac{1}{q} \int_{\mathbb{R}^N} b(\mathbf{x}) |u(\mathbf{x}/\sigma)|^q \sigma^{-\frac{qN}{p}}\, d\mathbf{x}$$

$$\leq \frac{1}{2} \int_{\mathbb{R}^N} |\nabla u_{\mathbf{x}/\sigma}(\mathbf{x}/\sigma)|^2 \sigma^{-\frac{2N}{p}-2} \sigma^N d(\mathbf{x}/\sigma) - \frac{1}{q} \int_{\mathbb{R}^N} \frac{C_1}{1+|\mathbf{x}|^\alpha} |u(\mathbf{x}/\sigma)|^q \sigma^{-\frac{qN}{p}}\, d\mathbf{x}$$

$$\leq \frac{1}{2} \sigma^{-\frac{2N}{p}-2+N} \|\nabla u\|_2^2$$

$$- \mathbf{C} \inf_{\mathbf{x} \in \mathbb{R}^N} \frac{1+|\mathbf{x}/\sigma|^\alpha}{1+|\mathbf{x}|^\alpha} \int_{\mathbb{R}^N} \frac{1}{1+|\mathbf{x}/\sigma|^\alpha} |u(\mathbf{x}/\sigma)|^q \sigma^{-\frac{qN}{p}} \sigma^N\, d(\mathbf{x}/\sigma)$$

$$\leq \sigma^{-\frac{2N}{p}-2+N} \left(\|\nabla u\|_2^2 - \mathbf{C}\, \sigma^{-\frac{N}{p}(q-2)+2-\alpha} \int_{\mathbb{R}^N} \frac{|u(\mathbf{y})|^q}{1+|\mathbf{y}|^\alpha}\, d\mathbf{y} \right), \tag{10.10}$$

where constants $\mathbf{C}$ do not depend on $\sigma \geq 1$. Because $-\frac{N}{p}(q-2)+2-\alpha > 0$, the right-hand side of (10.10) becomes negative for sufficiently large σ or, in other words, we may find a function u_σ with

$$J_1(u_\sigma) < 0, \qquad J_2(u_\sigma) = 1,$$

which proves (10.9). □

LEMMA 10.9. *For any $u \in \mathcal{E}_{p,q(b)}$ satisfying*

$$J_1(u) < 0, \tag{10.11}$$

$$J_2(u) < 1 \tag{10.12}$$

there exists $v \in \mathcal{E}_{p,q(b)}$ such that

$$J_1(v) < J_1(u), \qquad J_2(v) = 1.$$

Proof. Because $J_1(u) < 0$, u is nontrivial, therefore $J_2(u) \neq 0$. Suppose $v = ut$, where $t \in \mathbb{R}^+$ is defined as

$$t = (J_2(u))^{-1/p}.$$

We see that $t > 1$ by (10.12). Obviously,

$$J_2(v) = J_2(u)(J_2(u))^{-1} = 1,$$

and from $t > 1$ and $J_1(u) < 0$,

$$\begin{aligned} J_1(v) &= \frac{1}{2}t^2\|\nabla u\|_2^2 - \frac{t^q}{q}\|b^{1/q}u\|_q^q \\ &< t^2\left(\frac{1}{2}\|\nabla u\|_2^2 - \frac{1}{q}\|b^{1/q}u\|_q^q\right) = t^2 J_1(u) < J_1(u). \end{aligned}$$ □

Now we are ready to prove condition (IV).

LEMMA 10.10. *Condition (IV) holds.*

Proof. According to Lemma 10.8, estimate (10.9) is valid, thus we may choose $\varepsilon = -I/2 > 0$. Let for some $u \in \mathcal{E}_{p,q(b)}$,

$$J_1(u) - I < \varepsilon, \qquad J_2(u) < 1\,.$$

Then Lemma 10.9 becomes the statement of (IV). □

10.5 COMPLETING THE PROOF

All the conditions of Theorem 6.7 hold and consequently there exist λ_1 and λ_2 not simultaneously equal to zero and $u \in \mathcal{E}_{p,q(b)}$ such that

$$\lambda_1 J_1'(u) + \lambda_2 J_2'(u) = 0. \tag{10.13}$$

To finish the proof of Theorem 10.1 we need to show that $\lambda = -\lambda_2/\lambda_1 < 0$. Indeed, (10.13) implies

$$\lambda_1\langle J_1'(u), u\rangle + \lambda_2\langle J_2'(u), u\rangle = 0. \tag{10.14}$$

Because

$$\langle J_2'(u), u\rangle = pJ_2(u) = p,$$

we have by (10.13) and (10.14) that $\lambda_1 \neq 0$.

From Lemma 10.8 and relations (10.13) and (10.14) we get

$$\begin{aligned} 0 > I = J_1(u) &= \frac{1}{2}\|\nabla u\|_2^2 - \frac{1}{q}\|b^{1/q}u\|_q^q = \frac{1}{2}\langle J_1'(u), u\rangle + \left(\frac{1}{2} - \frac{1}{q}\right)\|b^{1/q}u\|_q^q \\ &= \frac{\lambda_2}{2\,\lambda_1}\langle J_2'(u), u\rangle + \left(\frac{1}{2} - \frac{1}{q}\right)\|b^{1/q}u\|_q^q \geq \frac{p\lambda_2}{2\,\lambda_1}. \end{aligned}$$

§11. The Comparison Method for Eigenvalue Problems (Concentration Compactness)

In this section we consider elliptic problems in the situation where appropriate theorems of §§7–10 which we used earlier do not work, and only Theorem 6.10 may be applied.

The main object of this section is disclosing the role of the method of comparing two variational problems in applications to the elliptic equations on $\mathbb{R}^N$, which is known as the concentration compactness method. Here we present the first step in this approach demonstrated on a simple example.

The pioneer elaboration of this method belongs to Lions [2,3]. Note that a modification of this approach is also applicable to problems arising in the study of equations with the critical exponent of nonlinearity (Brezis and Nirenberg [1]).

11.1 SETTING UP THE PROBLEM

Consider a problem of the form

$$-\Delta u + a(\mathbf{x})u - \lambda b(\mathbf{x})|u|^{q-2}u = 0 \qquad \text{in } \mathbb{R}^N, \tag{11.1}$$

where $2 < q < 2^*$, $\lambda > 0$, and $a, b : \mathbb{R}^N \to \mathbb{R}$ are measurable locally bounded nonnegative functions.

The peculiarity of the conditions in this paragraph is the assumption that there are a continuous nondecreasing function $\bar{a} : \mathbb{R}^+ \to \mathbb{R}^+$ and a positive constant $\bar{b}$ such that

(a)
$$\varliminf_{R\to\infty} \bar{a}(|\mathbf{x}|) > 0,$$

(b)
$$\lim_{R\to\infty} \sup_{\mathbf{x}\in\mathbb{R}^N\setminus B_R} \frac{|a(\mathbf{x}) - \bar{a}(|\mathbf{x}|)|}{\bar{a}(|\mathbf{x}|)} = 0,$$

(c)
$$\lim_{R\to\infty} \sup_{\mathbf{x}\in\mathbb{R}^N\setminus B_R} |b(\mathbf{x}) - \bar{b}| = 0.$$

In accordance with the method described in §6 we associate with problem (11.1) the variational problem

$$\{u \in \mathcal{E}_{2(a),q(b)}, \min J_1(u), J_2(u) = 1\} \tag{11.2}$$

with the functionals $J_1, J_2 : \mathcal{E}_{2(a),q(b)} \to \mathbb{R}$,

$$J_1(u) = \|\nabla u\|_2^2 + \|a^{1/2}u\|_2^2,$$
$$J_2(u) = \|b^{1/q}u\|_q^q.$$

Generally, the conditions imposed are not sufficient for the variational problem to have a solution. Let us consider a typical example without solution.

11.2 A NONEXISTENCE RESULT FOR THE VARIATIONAL PROBLEM

Suppose that $\bar{a} \equiv 1$, $a \in L_\infty(\mathbb{R}^N)$, $\bar{b} \equiv b \equiv 1$, and

$$a(\mathbf{x}) > 1$$

for all $\mathbf{x} \in \mathbb{R}^N$.

THEOREM 11.1. *Under conditions (a), (b), (c), and the above assumptions problem (11.2) has no solution.*

Proof. From conditions on b, $\mathcal{E}_{2(a),q(b)} = \mathcal{E}_{2(a),q}$. Assume a solution $u \in \mathcal{E}_{2(a),q}$ exists. Consider a sequence $\{u_n\} \subset \mathcal{E}_{2(a),q}$ defined by

$$u_n(\mathbf{x}) = u(\mathbf{x} + \mathbf{e}\,n) \tag{11.3}$$

where $\mathbf{e}$ is an ort in $\mathbb{R}^N$.

From the definition of J_2 and (11.3),

$$J_2(u_n) = \|u_n\|_q^q = \|u\|_q^q = J_2(u). \tag{11.4}$$

Now we consider how $J_1(u_n)$ behaves as $n \to \infty$. We have

$$\begin{aligned} J_1(u_n) &= \|\nabla u_n\|_2^2 + \int_{\mathbb{R}^N} a(\mathbf{x})|u_n|^2\,d\mathbf{x} \\ &= \|\nabla u\|_2^2 + \int_{\mathbb{R}^N} a(\mathbf{x})|u|^2\,d\mathbf{x} + \int_{\mathbb{R}^N} a(\mathbf{x})(|u_n|^2 - |u|^2)\,d\mathbf{x} \\ &= J_1(u) + \int_{\mathbb{R}^N} a(\mathbf{x})(|u_n|^2 - |u|^2)\,d\mathbf{x}. \end{aligned} \tag{11.5}$$

We estimate the right-hand side of (11.5). From (a) with $\bar{a} \equiv 1$ and (11.3), for any $\varepsilon > 0$ there exists a radius R_ε such that for all n,

$$\int\limits_{\mathbb{R}^N \setminus B_{R_\varepsilon}} (a(\mathbf{x}) - 1)|u_n|^2 \, d\mathbf{x} \leq \varepsilon \int\limits_{\mathbb{R}^N \setminus B_{R_\varepsilon}} |u_n|^2 \, d\mathbf{x} \leq \varepsilon \|u\|_2^2. \tag{11.6}$$

Because $u \in L_2(\mathbb{R}^N)$, for the ε, R_ε chosen there exists a number n_ε such that for all $n > n_\varepsilon$,

$$\int\limits_{B_{R_\varepsilon}} |u_n|^2 \, d\mathbf{x} = \int\limits_{n\mathbf{e}+B_{R_\varepsilon}} |u|^2 \, d\mathbf{x} \leq \varepsilon. \tag{11.7}$$

Combining (11.6) and (11.7) and passing to the limit $\varepsilon \to 0$, we get

$$\lim_{n\to\infty} \int\limits_{\mathbb{R}^N} a(\mathbf{x})|u_n|^2 \, d\mathbf{x} = \int\limits_{\mathbb{R}^N} |u|^2 \, d\mathbf{x}. \tag{11.8}$$

In other words, thanks to (11.5) and (11.8), we have the estimate

$$\lim_{n\to\infty} J_1(u_n) = J_1(u) + \int\limits_{\mathbb{R}^N} (|u|^2 - a(\mathbf{x})|u|^2) \, d\mathbf{x} < J_1(u).$$

This estimate, together with (11.4), contradicts the assumption that u solves (11.2), which completes the proof. □

11.3 AN EXISTENCE THEOREM FOR THE VARIATIONAL PROBLEM

Despite Section 11.2 we may formulate conditions under which problem 11.2 has a solution. Denote

$$I = \inf_{\substack{u \in \mathcal{E}_{2(a),q} \\ J_2(u)=1}} J_1(u), \qquad I^\infty = \inf_{\substack{u \in \mathcal{E}_{2(a),q} \\ J_2^\infty(u)=1}} J_1^\infty(u),$$

where

$$J_1^\infty(u) = \|\nabla u\|_2^2 + \|\bar{a}^{1/2} u\|_2^2,$$
$$J_2^\infty(u) = \|\bar{b}^{1/q} u\|_q^q.$$

Remark. If functions a, b satisfy the conditions of Section 11.2, then

$$I = I^\infty. \tag{11.9}$$

Indeed, $I \geq I^\infty$ because $a > \bar{a}$. The inverse inequality is true because for any $u \in \mathcal{E}_{2(a),q}$ and any sequence $\{u_n\}$ defined in accordance with (11.3), the following holds analogously to (11.4) and (11.8)

$$\lim_{n\to\infty} J_1(u_n) = J_1^\infty(u),$$
$$J_2(u_n) = J_2^\infty(u) \qquad \text{for all } n,$$

and (11.9) follows.

THEOREM 11.2. *Let $N > 2$, $2 < q < 2^*$, and conditions (a), (b), (c) hold, and let*

$$0 < I < I^\infty. \tag{11.10}$$

Then problem (11.2) has a solution.

We shall make use of Theorem 6.8 for proving Theorem 11.2. For this we need to verify its assumptions. We begin with the condition of nonemptiness of a corresponding set.

LEMMA 11.3. *The set*

$$\{u \in \mathcal{E}_{2(a),q}, J_2(u) = 1\}$$

is not empty.

Proof. See Lemma 7.1. □

LEMMA 11.4. *The functional J_1 is bounded from below, coercive on $\mathcal{E}_{2(a),q}$ under the condition $J_2 = 1$, and also weakly lower semicontinuous on this space.*

Proof. See Lemmas 7.2, 7.3, and 7.4. □

We now need only to verify condition (V) of Theorem 6.8. To do this, we consider some preliminary lemmas.

11.4 PRELIMINARY LEMMAS

LEMMA 11.5. *Let for a sequence $\{u_n\} \subset \mathcal{E}_{2(a),q}$,*

$$u_n \rightharpoonup u \qquad \text{in } \mathcal{E}_{2(a),q}. \tag{11.11}$$

Then

$$\lim_{n\to\infty} (\|b^{1/q}u_n\|_q^q - \|b^{1/q}(u_n - u)\|_q^q) = \|b^{1/q}u\|_q^q. \tag{11.12}$$

Proof. Clearly,

$$\int\limits_{\mathbb{R}^N\setminus B_R} b(\mathbf{x})|u_n|^q\,d\mathbf{x}$$

$$= \int\limits_{\mathbb{R}^N\setminus B_R} b(\mathbf{x})|u_n - u|^q\,d\mathbf{x} + q \int\limits_{\mathbb{R}^N\setminus B_R} b(\mathbf{x}) \int_0^1 |u_n - u + \theta u|^{q-2}(u_n - u + \theta u)u\,d\theta\,d\mathbf{x}.$$

Consequently,

$$\int_{\mathbb{R}^N} b(\mathbf{x})(|u_n|^q - |u|^q - |u_n - u|^q)\, d\mathbf{x}$$

$$= \int_{B_R} b(\mathbf{x})\,(|u_n|^q - |u|^q)\, d\mathbf{x} - \int_{B_R} b(\mathbf{x})\,|u_n - u|^q\, d\mathbf{x} - \int_{\mathbb{R}^N\setminus B_R} b(\mathbf{x})\,|u|^q\, d\mathbf{x}$$

$$+ q \int_{\mathbb{R}^N\setminus B_R} \int_0^1 b(\mathbf{x})\,|u_n - u + \theta u|^{q-2}(u_n - u + \theta u)u\, d\theta\, d\mathbf{x} \qquad (11.13)$$

From the Kondrashov theorem, (11.11) implies

$$u_n \to u \qquad \text{in } L_q(B_R)$$

for any R. Hence, from the Vainberg-Krasnoselskii theorem, the first and second integrals on the right-hand side of (11.13) tend to zero as $n \to \infty$ for any B_R. The third integral tends to zero as $R \to \infty$ because of the definition of $L_{q(b)}(\mathbb{R}^N)$. In order to estimate the last integral, we make use of the Hölder inequality and boundedness, from (11.11), of $\{u_n\}$ in $\mathcal{E}_{2(a),q}$. We get

$$\int_{\mathbb{R}^N\setminus B_R} \int_0^1 \left|b(\mathbf{x})\,|u_n - u + \theta u|^{q-2}(u_n - u + \theta u)u\right|\, d\theta\, d\mathbf{x}$$

$$\le \max_{0\le\theta\le1} \|b^{1/q}(u_n - u + \theta u)\|_q^{q-1}(\mathbb{R}^N \setminus B_R)\, \|b^{1/q}u\|_q(\mathbb{R}^N \setminus B_R)$$

$$\le \mathbf{C}\,(\|b^{1/q}u_n\|_q^{q-1} + \|b^{1/q}u\|_q^{q-1})\, \|b^{1/q}u\|_q(\mathbb{R}^N \setminus B_R)$$

$$\le \mathbf{C}\, \|b^{1/q}u\|_q(\mathbb{R}^N \setminus B_R).$$

where constants $\mathbf{C}$ do not depend on n, R. This means that the fourth integral on the right-hand side of (11.13) converges to zero as $R \to \infty$ uniformly in n.

Thus, choosing an arbitrary $\varepsilon > 0$, we may find R_ε such that the absolute value of the third integral is less than ε and the same is true for the fourth integral for all n. Given ε and R_ε, we find a number n_ε such that, for all $n \ge n_\varepsilon$, absolute values of the first and second integrals are less than ε. After taking the limit $\varepsilon \to 0$, we get (11.12). □

LEMMA 11.6. *Let for a sequence* $\{u_n\} \subset \mathcal{E}_{2(a),q}$,

$$u_n \rightharpoonup 0 \qquad \textit{in } \mathcal{E}_{2(a),q}. \qquad (11.14)$$

Then

$$\begin{aligned} \lim_{n\to\infty} (\|a^{1/2}u_n\|_2^2 - \|\bar{a}^{1/2}u_n\|_2^2) &=0, \\ \lim_{n\to\infty} (\|b^{1/q}u_n\|_q^q - \|\bar{b}^{1/q}u_n\|_q^q) &=0. \end{aligned} \tag{11.15}$$

Proof. We shall prove the first relation of (11.15). The second may be proved analogously. From (a), for any $\varepsilon > 0$ there exists R_ε such that

$$\sup_{\mathbf{x}\in\mathbb{R}^N\setminus B_R} \frac{|a(\mathbf{x}) - \bar{a}(|\mathbf{x}|)|}{\bar{a}(|\mathbf{x}|)} \le \varepsilon/2,$$

that is, for this radius, all u_n, and sufficiently small ε,

$$\int\limits_{\mathbb{R}^N\setminus B_R} |a(\mathbf{x}) - \bar{a}(|\mathbf{x}|)|\,|u_n|^2\,d\mathbf{x} \le \varepsilon \int\limits_{\mathbb{R}^N\setminus B_R} a(|\mathbf{x}|)\,|u_n|^2\,d\mathbf{x} \le \varepsilon\|a^{1/2}u_n\|_2^2. \tag{11.16}$$

On the other hand, from the Kondrashov theorem, (11.14) implies

$$u_n \to 0 \qquad \text{in } L_q(B_R) \tag{11.17}$$

for any B_R. At last, by (11.14),

$$\|a^{1/2}u_n\|_2 \le \mathbf{C}.$$

Now, given ε, we may by (11.16), choose R_ε, and, given ε and R_ε, we may by (11.17), choose a number n_ε such that for all $n \ge n_\varepsilon$ the estimate

$$\begin{aligned} \int\limits_{\mathbb{R}^N} \big|a(\mathbf{x})\,|u_n|^2 - \bar{a}(|\mathbf{x}|)\,|u_n|^2\big|\,d\mathbf{x} &= \int\limits_{\mathbb{R}^N\setminus B_{R_\varepsilon}} |a(\mathbf{x}) - \bar{a}(|\mathbf{x}|)|\,|u_n|^2\,d\mathbf{x} \\ &+ \int\limits_{B_{R_\varepsilon}} |a(\mathbf{x}) - \bar{a}(|\mathbf{x}|)|\,|u_n|^2\,d\mathbf{x} \le \varepsilon\,\|a^{1/2}u_n\|_2^2 + \varepsilon \le \varepsilon\mathbf{C} \end{aligned}$$

holds, where $\mathbf{C}$ does not depend on both n and ε. Taking the limit $\varepsilon \to 0$, we get the result. □

11.5 CONDITION (V)

At last, we are ready to verify condition (V).

LEMMA 11.7. *For any sequence* $\{u_n\} \subset \mathcal{E}_{2(a),q}$ *satisfying*

$$u_n \rightharpoonup u \qquad \text{in } \mathcal{E}_{2(a),q},$$
$$J_1(u_n) \to I, \tag{11.18}$$
$$J_2(u_n) = 1, \tag{11.19}$$

where I *is defined in Section 11.3,*

$$J_2(u) = 1$$

holds.

Proof. This lemma follows from the continuousness of J_2 on $\mathcal{E}_{2(a),q}$ and the following assertion:

LEMMA 11.8 (ON CONDITIONAL COMPACTNESS). *Any sequence satisfying the conditions of Lemma 11.7 converges strongly in* $\mathcal{E}_{2(a),q}$.

Proof. Let

$$u_n \rightharpoonup u \qquad \text{in } \mathcal{E}_{2(a),q}.$$

Denote

$$v_n = u_n - u.$$

Obviously,

$$v_n \rightharpoonup 0 \qquad \text{in } \mathcal{E}_{2(a),q}.$$

All we need is to show that

$$v_n \to 0 \qquad \text{in } \mathcal{E}_{2(a),q}. \tag{11.20}$$

Indeed, because of the weak convergence, we have

$$\begin{aligned}\lim_{n\to\infty} (\|\nabla u_n\|_2^2 - \|\nabla v_n\|_2^2) &= \|\nabla u\|_2^2, \\ \lim_{n\to\infty} (\|a^{1/2}u_n\|_2^2 - \|a^{1/2}v_n\|_2^2) &= \|a^{1/2}u\|_2^2.\end{aligned} \tag{11.21}$$

From Lemma 11.5, we get also

$$\lim_{n\to\infty} (\|b^{1/q}u_n\|_q^q - \|b^{1/q}v_n\|_q^q) = \|b^{1/q}u\|_q^q. \tag{11.22}$$

For any two positive numbers x, y, and $z > 1$ the inequality

$$(x+y)^{1/z} \le x^{1/z} + y^{1/z}$$

is true, therefore (11.22) implies the existence of a subsequence such that

$$\lim_{n\to\infty}(\|b^{1/q}u_n\|_q^2 - \|b^{1/q}v_n\|_q^2) \le \|b^{1/q}u\|_q^2. \tag{11.23}$$

From the definition of I and (11.18), (11.19), we have

$$\lim_{n\to\infty}\frac{\|\nabla u_n\|_2^2 + \|a^{1/2}u_n\|_2^2}{\|b^{1/q}u_n\|_q^2} = I. \tag{11.24}$$

Besides, for any $u \in \mathcal{E}_{2(a),q}$ we have, by the definition of I,

$$\frac{\|\nabla u\|_2^2 + \|a^{1/2}u\|_2^2}{\|b^{1/q}u\|_q^2} \ge I. \tag{11.25}$$

Passing to the limit in accordance with (11.21) and (11.23), and taking into consideration (11.24) and (11.25), we get the estimate

$$\begin{aligned}\lim_{n\to\infty}&(\|\nabla v_n\|_2^2 + \|a^{1/2}v_n\|_2^2 - I\|b^{1/q}v_n\|_q^2)\\ &\stackrel{21,23}{\le} \lim_{n\to\infty}(\|\nabla u_n\|_2^2 + \|a^{1/2}u_n\|_2^2 - I\|b^{1/q}u_n\|_q^2)\\ &\qquad - (\|\nabla u\|_2^2 + \|a^{1/2}u\|_2^2 - I\|b^{1/q}u\|_q^2) \stackrel{24,25}{\le} 0.\end{aligned} \tag{11.26}$$

By the definition of I^∞ we have the estimate

$$\frac{\|\nabla v_n\|_2^2 + \|\bar{a}^{1/2}v_n\|_2^2}{\|\bar{b}^{1/q}v_n\|_q^2} \ge I^\infty \tag{11.27}$$

for all $v_n \in \mathcal{E}_{2(a),q}$. On the other hand, by Lemma 11.6, the relations

$$\begin{aligned}&\lim_{n\to\infty}(\|a^{1/2}v_n\|_2^2 - \|\bar{a}^{1/2}v_n\|_2^2) = 0,\\ &\lim_{n\to\infty}(\|b^{1/q}v_n\|_q^q - \|\bar{b}^{1/q}v_n\|_q^q) = 0\end{aligned} \tag{11.28}$$

hold. Consequently, passing to the limit according to (11.28) and using estimates (11.27) and (11.26), we get

$$\begin{aligned}\lim_{n\to\infty}&\left(1-\frac{I}{I^\infty}\right)(\|\nabla v_n\|_2^2 + \|\bar{a}^{1/2}v_n\|_2^2)\\ &\stackrel{27}{\le} \lim_{n\to\infty}(\|\nabla v_n\|_2^2 + \|\bar{a}^{1/2}v_n\|_2^2 - I\|\bar{b}^{1/q}v_n\|_q^2)\\ &\qquad \stackrel{28}{=} \lim_{n\to\infty}(\|\nabla v_n\|_2^2 + \|a^{1/2}v_n\|_2^2 - I\|b^{1/q}v_n\|_q^2) \stackrel{26}{\le} 0.\end{aligned}$$

Hence, from (11.10), we have

$$\lim_{n\to\infty}(\|\nabla v_n\|_2^2 + \|\bar{a}^{1/2}v_n\|_2^2) \le 0,$$

which indeed is statement (11.20). □

Thus, Lemma 11.8 and, consequently, Lemma 11.7 are proved. □

All the assumptions of Theorem 6.8 are valid, hence Theorem 11.2 is true. □

In accordance with Theorem 6.9 the following also holds:

THEOREM 11.9. *Under the conditions of Theorem 11.2 there exists $\lambda > 0$ such that problem (11.1) has a nontrivial solution $u \in \mathcal{E}_{2(a),q}$.*

11.6 AN EXAMPLE

Condition (11.10) does not have an explicit form. It would be better to have direct conditions on functions a and b. Recall that we are considering the problem

$$-\Delta u + a(\mathbf{x})\, u - \lambda\, b(\mathbf{x})\, |u|^{q-2}u = 0 \qquad \text{in } \mathbb{R}^N, \tag{11.29}$$

where $N > 2$, $2 < q < 2^*$, and a, b are measurable locally bounded functions.

THEOREM 11.10. *Let conditions (a), (b), (c) of §11 hold. Let*

$$a \le \bar{a} \quad \text{and} \quad b \ge \bar{b}.$$

Then there exists $\lambda > 0$ such that problem (11.29) has a nontrivial solution $u \in \mathcal{E}_{2(\bar{a}),q}$.

Proof. Let $\mathcal{E}^{\mathrm{rad}}_{2(\bar{a}),q}$ be the subspace of $\mathcal{E}_{2(a),q}$ consisting of radial functions. Due to the Schwarz symmetrization (see Appendix D, Theorem D.4), we have

$$\inf_{\substack{u\in\mathcal{E}_{2(\bar{a}),q}\\ J_2^\infty=1}} J_1^\infty = \inf_{\substack{u\in\mathcal{E}^{\mathrm{rad}}_{2(\bar{a}),q}\\ J_2^\infty=1}} J_1^\infty.$$

The problem

$$\left\{u \in \mathcal{E}^{\mathrm{rad}}_{2(\bar{a}),q},\ \min J_1^\infty(u),\ J_2^\infty(u) = 1\right\}$$

was studied in §8. There the existence of its nontrivial solution satisfying the equation

$$-\Delta u + \bar{a}(|\mathbf{x}|)\, u - \lambda\, \bar{b}\, |u|^{q-2}u = 0 \qquad \text{on } \mathbb{R}^N \tag{11.30}$$

with $\lambda > 0$ was proved. Denote

$$\sigma = \|b^{1/q}u\|_q^q$$

and define $w \in \mathcal{E}_{2(a),q}$ by

$$w = \sigma^{-1/q}u.$$

It is obvious that

$$\sigma \ge \|\bar{b}^{1/q}u\|_q^q = 1.$$

Then

$$J_2(w) = 1$$

and

$$\begin{aligned} I \le J_1(w) &= \|\nabla w\|_2^2 + \|a^{1/2}w\|_2^2 \le \|\nabla w\|_2^2 + \|\bar{a}^{1/2}w\|_2^2 \\ &= \sigma^{-2/q}(\|\nabla u\|_2^2 + \|\bar{a}^{1/2}u\|_2^2) = \sigma^{-2/q}J_1^\infty(u) = \sigma^{-2/q}I^\infty \le I^\infty. \end{aligned} \tag{11.31}$$

In accordance with (11.31) $I \le I^\infty$. Now two variants are possible. If $I = I^\infty$, then, by (11.31), the definition of σ, and conditions on a, b,

$$\bar{a}(|\mathbf{x}|) - a(\mathbf{x}) \equiv \bar{b}(|\mathbf{x}|) - b(\mathbf{x}) \equiv 0,$$

and hence u satisfies (11.30) with corresponding λ. If $I < I^\infty$, then Theorem 11.2 implies Theorem 11.10. □

§12. Homogeneous Problems

In this section we consider problems of the form

$$-\Delta u + f(\mathbf{x}, u) = 0 \qquad \text{in } \mathbb{R}^N, \tag{12.1}$$

where as compared with the eigenvalue problems, a free parameter λ is absent.

It occurs that if f possesses some homogeneity properties, the existence of a nontrivial solution to (12.1) follows from a theorem on existence of a solution to the corresponding eigenvalue problem. We consider two such examples.

12.1 The First Example

Let f have the form

$$f(\mathbf{x}, t) = a(\mathbf{x})\, t - b(\mathbf{x})\, |t|^{q-2} t, \qquad q > 1.$$

It is easy to see that if for some $\lambda > 0$, the problem

$$-\Delta u + a(\mathbf{x})\, u - \lambda b(\mathbf{x}) |u|^{q-2} u = 0 \qquad \text{in } \mathbb{R}^N,$$

has a nontrivial solution belonging to one of the spaces considered earlier, then the function

$$v = \lambda^{1/(q-2)} u$$

is a nontrivial solution to

$$-\Delta v + a(\mathbf{x})\, v - b(\mathbf{x}) |v|^{q-2} v = 0 \qquad \text{in } \mathbb{R}^N. \tag{12.2}$$

This means that all the existence theorems applied to the corresponding eigenvalue problem may be reformulated in terms of problem (12.2).

12.2 The Second Example

Let f have the form

$$f(\mathbf{x}, t) = f(t),$$

i.e., f does not depend on $\mathbf{x}$. It is easy to see that if for some $\lambda > 0$ the problem

$$-\Delta u + \lambda f(u) = 0 \qquad \text{in } \mathbb{R}^N,$$

has a nontrivial solution corresponding to one of the spaces considered earlier, then

$$v(\,.\,) = u(\,.\,\lambda^{-1/2})$$

is a nontrivial solution to

$$-\Delta v + f(v) = 0 \qquad \text{in } \mathbb{R}^N. \tag{12.3}$$

This means that all the existence theorems applied to the corresponding eigenvalue problem may be reformulated in terms of problem (12.3).

Chapter 3
Special Variational Methods

§13. The Mountain Pass Method

We saw in §12 that noncoercive problems

$$
\begin{aligned}
& -\Delta u + f(\mathbf{x}, u) = 0 \qquad \text{in } \mathbb{R}^N, \\
& u \to 0 \qquad \text{as } |\mathbf{x}| \to \infty,
\end{aligned}
\tag{13.1}
$$

can be studied with the help of theorems on a conditional minimum if f is homogeneous. However such f are rare.

Nevertheless, there are methods permitting the study of noncoercive boundary problems for an equation of type (13.1) without an indefinite parameter λ. One of these methods utilizes the so-called mountain pass theorem of Ambrosetti and Rabinowitz [1]. In this section this theorem will be given in an abstract form (for the proof see Appendix E) and then it will be applied to an appropriate elliptic boundary problem.

13.1 Formulation of the Mountain Pass Theorem

Let X be a Banach space and $J : X \to \mathbb{R}$ be a functional of the class C^1.

Let θ be the origin of X and $y \neq \theta$ be some other of its elements. Let $\mathcal{P}$ be a family of paths π connecting the points θ and y, i.e., a family of continuous mappings $\pi : [0,1] \to X$ with

$$\pi(0) = \theta, \qquad \pi(1) = y.$$

Now we formulate a variant of the mountain pass theorem.

THEOREM 13.1. *Let U be a neighborhood of θ with $y \notin U$ such that*

$$\max\{J(\theta),\ J(y)\} < \inf_{u \in \partial U} J(u). \tag{13.2}$$

Then there exists a sequence $\{u_n\} \subset X$ such that

$$
\begin{aligned}
J(u_n) &\to I \qquad \text{in } \mathbb{R}, \\
J'(u_n) &\to 0 \qquad \text{in } X^*
\end{aligned}
\tag{13.3}
$$

as $n \to \infty$, where I is defined by

$$I = \inf_{\pi \in \mathcal{P}} \sup_{t \in [0,1]} J(\pi(t)). \tag{13.4}$$

DEFINITION OF PS-SEQUENCES. A sequence in X satisfying for some $c \in \mathbb{R}$

$$\begin{aligned} J(u_n) &\to c \qquad \text{in } \mathbb{R}, \\ J'(u_n) &\to 0 \qquad \text{in } X^* \end{aligned}$$

is called a PS$_c$-sequence (Palais-Smale sequence).

The properties of PS$_c$-sequences are important for the study of differential equations.

DEFINITION OF THE PS$_c$-CONDITION. A functional J of the class $C^1(X)$ is said to be satisfying the PS$_c$-condition on the Banach space X if one can extract from any corresponding PS$_c$-sequence $\{u_n\}$ a subsequence strongly convergent in X to some of its elements.

Remark. If c is not specified we omit it in the notation.

For such functionals the following variant of the mountain pass theorem is true:

THEOREM 13.2. *Let (13.2) be satisfied. Let a functional J satisfy PS_I on X where I is defined by (13.4). Then there is a $u \in X$ such that*

$$\begin{aligned} J(u) &= I \qquad \textit{in } \mathbb{R}, \\ J'(u) &= 0 \qquad \textit{in } X^*. \end{aligned}$$

Proof. Theorems 13.1 and 13.2 are proved in Appendix E. □

13.2 SETTING UP THE BOUNDARY PROBLEM

We show how the general Theorems 13.1 and 13.2 may be used for studying nontrivial solvability of noncoercive elliptic equations in $\mathbb{R}^N$. Consider the problem

$$-\Delta u + a(\mathbf{x})|u|^{p-2}u - b(\mathbf{x})|u|^{q-2}u = 0 \qquad \text{in } \mathbb{R}^N, \tag{13.5}$$

where $N > 2$ and $2 \leq p < q < 2^*$.

THEOREM 13.3. *Let nontrivial nonnegative measurable a and b satisfy assumptions of one of the Theorems 7.10, 7.11, or 7.12. Then problem (13.5) has a nontrivial solution $u \in \mathcal{E}_{p(a),q(b)}$.*

Proof. We shall use the mountain pass theorem. Choose $\mathcal{E}_{p(a),q(b)}$ as the space X. As the functional J consider the problem functional Φ for (13.5):

$$\Phi(u) = \frac{1}{2}\|\nabla u\|_2^2 + \frac{1}{p}\|a^{1/p}u\|_p^p - \frac{1}{q}\|b^{1/q}u\|_q^q. \tag{13.6}$$

To move forward we use Theorem 13.2.

13.3 VERIFYING ASSUMPTION (13.2)

The checking of (13.2) is based on the following three lemmas.

LEMMA 13.4. $\Phi(0) = 0$.

Proof. The proof is obvious in view of definition (13.6). □

LEMMA 13.5. *There exists a neighborhood U of θ such that*

$$\inf_{u \in \partial U} \Phi(u) > 0.$$

Proof. Define a neighborhood U^r by $\|\nabla u\|_2 + \|a^{1/p}u\|_p \le r$. Taking into consideration the corresponding embedding Theorem 7.10, 7.11, or 7.12, we have for $u \in U^r$

$$\mathbf{C}_1 r^q \ge \mathbf{C}_1 \left(\|\nabla u\|_2 + \|a^{1/p}u\|_p\right)^q \ge \|b^{1/q}u\|_q^q,$$

where $\mathbf{C}_1$ does not depend on u and r. It easy to see that for $u \in \partial U^r$ with $0 < r \le 1$,

$$\frac{1}{2}\|\nabla u\|_2^2 + \frac{1}{p}\|a^{1/p}u\|_p^p \ge \frac{1}{p}(\|\nabla u\|_2^p + \|a^{1/p}u\|_p^p) \ge \frac{\mathbf{C}_2}{p}(\|\nabla u\|_2 + \|a^{1/p}u\|_p)^p = \frac{\mathbf{C}_2}{p} r^p,$$

where $\mathbf{C}_1 \ge \mathbf{C}_2 > 0$ do not depend on u and r.

Hence, for $0 < r \le 1$,

$$\inf_{u \in \partial U^r} \Phi(u) \ge \frac{1}{p}\mathbf{C}_2 r^p - \frac{1}{q}\mathbf{C}_1 r^q.$$

Choose

$$r_0 = \left(\frac{\mathbf{C}_2}{\mathbf{C}_1}\right)^{\frac{1}{q-p}} \le 1.$$

Simple arithmetic gives the estimate

$$\inf_{u \in \partial U^{r_0}} \Phi(u) \ge \left(\frac{1}{p} - \frac{1}{q}\right)\left(\frac{\mathbf{C}_2^q}{\mathbf{C}_1^p}\right)^{\frac{1}{q-p}} > 0.$$

Thus, we can choose $U = U^{r_0}$ as a suitable neighborhood. □

LEMMA 13.6. *There is a $y_0 \in X$ such that*

$$y_0 \notin U,$$
$$\Phi(y_0) < 0.$$

Proof. Choose some function $\varphi \in \mathcal{D}(\mathbb{R}^N)$ with nontrivial $b\varphi$ and consider a set of functions $\{\varphi_t\}$ of the form $\varphi_t = t\,\varphi$, $t \in \mathbb{R}^+$. By force of definition (13.6)

$$\Phi(\varphi_t) = \frac{1}{2}\|\nabla\varphi\|_2^2\, t^2 + \frac{1}{p}\|a^{1/p}\varphi\|_p^p\, t^p - \frac{1}{q}\|b^{1/q}\varphi\|_q^q\, t^q.$$

Since $q > p \geq 2$, there is a t_0 large enough such that

$$(\|\nabla\varphi\|_2 + \|a^{1/p}\varphi\|_p)\, t_0 > r_0,$$
$$\Phi(\varphi_{t_0}) < 0.$$

Choosing $y_0 = \varphi_{t_0}$, we prove the lemma. □

To complete the verifying of (13.2), consider a set $\mathcal{P}$ of paths $\pi : [0,1] \to \mathcal{E}_{p(a),q(b)}$ connecting the points $\theta \in \mathcal{E}_{p(a),q(b)}$ and $y_0 \in \mathcal{E}_{p(a),q(b)}$ from Lemma 13.6. Let U be U^{r_0} as in Lemma 13.5. Obviously, by Lemmas 13.4, 13.5, and 13.6,

$$\max\{\Phi(y_0), \Phi(\theta)\} = 0 < \inf_{u \in \partial U^{r_0}} \Phi(u)$$

and thus the proof is completed. □

13.4 VERIFYING THE PS-CONDITION. COMPLETING THE PROOF OF THEOREM 13.3

Verifying the PS-condition will be divided into a number of lemmas.

LEMMA 13.7. *From any PS-sequence of Φ on $\mathcal{E}_{p(a),q(b)}$ one can extract a subsequence weakly convergent to some element of the space $\mathcal{E}_{p(a),q(b)}$.*

Proof. Let $\{u_n\} \subset \mathcal{E}_{p(a),q(b)}$ be a PS-sequence of Φ, i.e.,

$$\begin{aligned} \Phi(u_n) &\to c \qquad \text{in } \mathbb{R}, \\ \Phi'(u_n) &\to 0 \qquad \text{in } \mathcal{E}^*_{p(a),q(b)} \end{aligned}$$

for some c. Choose arbitrarily $1/q < \theta < 1/p$. From definition (13.6) we have the following estimate:

$$\begin{aligned} &\left(\frac{1}{2} - \theta\right)\|\nabla u_n\|_2^2 + \left(\frac{1}{p} - \theta\right)\|a^{1/p}u\|_p^p \leq \left(\frac{1}{2} - \theta\right)\|\nabla u_n\|_2^2 \\ &+ \left(\frac{1}{p} - \theta\right)\|a^{1/p}u\|_p^p - \left(\frac{1}{q} - \theta\right)\|b^{1/q}u_n\|_q^q = \Phi(u_n) - \theta\,\langle \Phi'(u_n), u_n\rangle \qquad (13.7) \\ &= c + \mu_n + \langle \xi_n, u_n\rangle \leq \mathbf{C} + \|\xi_n\|_{\mathcal{E}^*_{p(a),q(b)}}\, \|u_n\|_{\mathcal{E}_{p(a),q(b)}} \\ &\leq \mathbf{C} + \mathbf{C}\,\|u_n\|_{\mathcal{E}^*_{p(a),q(b)}}, \end{aligned}$$

because $\mu_n \to 0$ in $\mathbb{R}$ and $\xi_n \to 0$ in $\mathcal{E}^*_{p(a),q(b)}$.

From (13.7), the sequence $\{u_n\}$ satisfies

$$\|\nabla u_n\|_2 + \|a^{1/p} u_n\|_p \le \mathbf{C}.$$

The corresponding embedding Theorem 7.10, 7.11, or 7.12 then provides

$$\|\nabla u_n\|_2 + \|a^{1/p} u_n\|_p + \|b^{1/q} u_n\|_q \le \mathbf{C}.$$

Because the space $\mathcal{E}_{p(a),q(b)}$ is reflexive, one can choose a weakly convergent subsequence. □

LEMMA 13.8. *Any sequence satisfying*

$$\Phi'(u_n) \to 0 \qquad \text{in } \mathcal{E}^*_{p(a),q(b)}, \tag{13.8}$$

$$u_n \rightharpoonup u \qquad \text{in } \mathcal{E}_{p(a),q(b)} \tag{13.9}$$

converges strongly to u *in* $\mathcal{E}_{p(a),q(b)}$.

Proof. Relation (13.9) gives

$$\|\nabla u_n\|_2 + \|a^{1/p} u_n\|_p + \|b^{1/q} u_n\|_q \le \mathbf{C}. \tag{13.10}$$

From (13.8) we have

$$|\langle \Phi'(u_n), u_n - u_m\rangle| \le \|\xi_n\|_{\mathcal{E}^*_{p(a),q(b)}} \|u_n - u_m\|_{\mathcal{E}_{p(a),q(b)}},$$
$$|\langle \Phi'(u_m), u_n - u_m\rangle| \le \|\xi_m\|_{\mathcal{E}^*_{p(a),q(b)}} \|u_n - u_m\|_{\mathcal{E}_{p(a),q(b)}},$$

where $\xi_n \to 0$ in $\mathcal{E}^*_{p(a),q(b)}$ as $n \to \infty$. Hence,

$$\begin{aligned}
&\Big| \|\nabla u_n - \nabla u_m\|_2^2 \\
&\qquad + \int_{\mathbb{R}^N} a(\mathbf{x}) \left(|u_n|^{p-1} u_n - |u_m|^{p-1} u_m\right)(u_n - u_m)\, d\mathbf{x} \\
&\qquad - \int_{\mathbb{R}^N} b(\mathbf{x}) \left(|u_n|^{q-1} u_n - |u_m|^{q-1} u_m\right)(u_n - u_m)\, d\mathbf{x} \Big| \\
&\qquad = |\langle \Phi'(u_n) - \Phi'(u_m), u_n - u_m\rangle| \\
&\qquad \le \left(\|\xi_n\|_{\mathcal{E}^*_{p(a),q(b)}} + \|\xi_m\|_{\mathcal{E}^*_{p(a),q(b)}}\right) \|u_n - u_m\|_{\mathcal{E}_{p(a),q(b)}} \\
&\qquad \le \mathbf{C}\left(\|\xi_n\|_{\mathcal{E}^*_{p(a),q(b)}} + \|\xi_m\|_{\mathcal{E}^*_{p(a),q(b)}}\right) \to 0 \qquad \text{as } m \ge n \to \infty.
\end{aligned} \tag{13.11}$$

By the Hölder inequality and the corresponding compact embedding theorem, (13.9) implies

$$\left| \int_{\mathbb{R}^N} b(\mathbf{x}) \left(|u_n|^{q-1}u_n - |u_m|^{q-1}u_m\right)(u_n - u_m)\, d\mathbf{x} \right|$$

$$\leq \|b^{1/q}(u_n - u_m)\|_q \left(\|b^{1/q}u_n\|_q^{q-1} + \|b^{1/q}u_m\|_q^{q-1}\right) \to 0 \quad (13.12)$$

as $m \geq n \to \infty$. Because $p \geq 2$,

$$\|a^{1/p}(u_n - u_m)\|_p^p \leq \mathbf{C} \int_{\mathbb{R}^N} a(\mathbf{x}) \left(|u_n|^{p-2}u_n - |u_m|^{p-2}u_m\right)(u_n - u_m)\, d\mathbf{x}. \quad (13.13)$$

Taking into consideration (13.11), (13.12), and (13.13) we get

$$\|\nabla u_n - \nabla u_m\|_2^2 + \|a^{1/p}(u_n - u_m)\|_p^p \to 0 \qquad \text{as } m \geq n \to \infty,$$

i.e., the sequence $\{u_n\}$ is a Cauchy sequence, hence it converges strongly to some $u_0 \in \mathcal{E}_{p(a),q(b)}$. Weak and strong limits coincide, therefore $u_0 = u$ and the lemma is proved. □

Completing the verifications of the PS-condition. By Lemma 13.7 and the definition of PS-sequences, the assumptions of Lemma 13.8 are valid. This lemma provides the validness of the PS-condition. □

Proof of Theorem 13.3. As we have seen, all the hypotheses of Theorem 13.2 are valid. Therefore equation (13.5) has a solution u in $\mathcal{E}_{p(a),q(b)}$: $\Phi'(u) = 0$ in $\mathcal{E}^*_{p(a),q(b)}$, with $\Phi(u) = I$. This solution is nontrivial because $\Phi(0) = 0$ and $I > 0$. Hence, Theorem 13.3 is proved. □

13.5 SIGN OF SOLUTIONS

We shall show that under the conditions of Theorem 13.3 one can establish analogously to Section 7.8 the existence of a nonnegative (or nonpositive) solution to (13.5).

THEOREM 13.9. *Let the conditions of Theorem 13.3 hold. Then problem (13.5) has a nonnegative (also nonpositive) solution.*

Proof. To prove this theorem we make use of the following statement.

THEOREM 13.10. *Let the conditions of Theorem 13.2 hold. Then for any sequence of paths* $\{\pi_n\}$ *satisfying*

$$\lim_{n\to\infty} \min_{u\in\pi_n} J(u) = I \tag{13.14}$$

there exists a PS_I*-sequence* $\{u_n\} \subset X$ *satisfying the relation*

$$\|u_n - \pi_n\|_X \to 0 \qquad \text{as } n \to \infty.$$

Proof. See Appendix E. □

Remark. A sequence of paths satisfying (13.14) will be called the minimizing path sequence.

Proof of Theorem 13.9. We establish the existence of a nonnegative solution. The case of a nonpositive solution is analogous.

It was mentioned in Section 7.8 that $u \in \mathcal{E}_{p(a),q(b)}$ implies $|u| \in \mathcal{E}_{p(a),q(b)}$. It follows also that $\Phi(|u|) = \Phi(u)$. This means that if $\{\pi_n(\,.\,)\} \subset \mathcal{P}$ is a minimizing paths sequence, so is $\{\pi_n(|\,.\,|)\} \subset \mathcal{P}$. By the definition of I a minimizing path sequence does exist. Hence, by Theorem 13.10, we may state the existence of such a PS_I-sequence $\{u_n\}$ for which there is a nonnegative sequence $\{v_n\}$ with $v_n \in \pi(|\,.\,|)$ and

$$\lim_{n\to\infty} \|u_n - v_n\|_{\mathcal{E}_{p(a),q(b)}} = 0. \tag{13.15}$$

In proving Theorem 13.3 we established the existence of a subsequence for any PS_c-sequence $\{u_n\}$ of the problem functional for (13.5), which converges strongly to a solution $u \in \mathcal{E}_{p(a),q(b)}$. By (13.15) we may find a sequence $\{v_n\}$ of nonnegative functions v_n which also converges to u. Hence u is nonnegative. □

13.6 A GENERAL THEOREM

Theorem 13.3 may be generalized in the case of problems

$$-\Delta u + f(\mathbf{x}, u) = 0 \qquad \text{in } \mathbb{R}^N, \tag{13.16}$$

where f is a Carathéodory function. Consider one of the possible variants. Denote

$$F(\mathbf{x}, t) = \int_0^t f(\mathbf{x}, s)\, ds.$$

(1) Let $f = f_1 + f_2$, where $f_1 \in K_{p(a),q(b)}$, $f_2 \in K_{q(b)}$ with $2 \le p < q < 2^*$, $N > 2$, and the measurable nonnegative functions a and b satisfy the conditions of one of Theorems 7.10, 7.11, or 7.12.

(2) Let there be a constant $\theta < 1/2$, positive constants γ_1, γ_2, γ_3, γ_4, and a function $g_0 \in L_1(\mathbb{R}^N)$ such that

$$F(\mathbf{x},t) - \theta\, f(\mathbf{x},t)\, t \geq \gamma_1\, a(\mathbf{x})\, |t|^p + g_0(\mathbf{x}),$$

$$(f_1(\mathbf{x},t') - f_1(\mathbf{x},t''))(t'-t'') \geq \gamma_2\, a(\mathbf{x})\, |t'-t''|^p,$$

$$F(\mathbf{x},t) \geq \gamma_3\, a(\mathbf{x})\, |t|^p - \gamma_4\, b(\mathbf{x})\, |t|^q$$

for almost all $\mathbf{x} \in \mathbb{R}^N$ and all $t, t', t'' \in \mathbb{R}$.

(3) Let there be $u_0 \in \mathcal{E}_{p(a),q(b)}$ such that

$$\frac{1}{2}\|\nabla u_0\|_2^2 + \int_{\mathbb{R}^N} F(\mathbf{x}, u_0)\, d\mathbf{x} < 0.$$

The following analog of Theorem 13.3 is true:

THEOREM 13.11. *Let all the assumptions of this section hold. Then equation (13.16) has a nontrivial solution $u \in \mathcal{E}_{p(a),q(b)}$.*

Proof. The proof is analogous to that of Theorem 13.3. □

Other variants of existence theorems may be found, for example, in the works of Noussair and Swanson [1,2].

§14. Behavior of PS-sequences. The Concentration Compactness (Comparison) Method

The aim of this section is to unify the ideas of the mountain pass theorem and the concentration compactness method, considered in §§11 and 13, so that the concentration compactness method can be applied to the problems without a parameter λ. In other words, we shall consider a boundary problem of the form

$$-\Delta u + f(\mathbf{x}, u) = 0 \qquad \text{in } \mathbb{R}^N, \tag{14.1}$$

as in the previous section. But in contrast to its assumptions (which resemble the assumptions of §7) we shall impose on the function f conditions similar to those considered in §11, that is, we shall be interested in the case

$$\lim_{|\mathbf{x}|\to\infty} f(\mathbf{x},t) = \bar{f}(t)$$

with some continuous $\bar{f}$.

Under these conditions difficulties similar to those encountered in §11 arise because the problem functional generally fails to satisfy the PS-condition. To study a more complicated behavior of PS-sequences, the approach based on the comparison of two variational problems (concentration compactness method) can be used.

Here we will study the behavior of PS-sequences for the problem functional of (14.1). The consequences of this result will be developed further.

We follow the pioneer works of Lions [1], and elaborations of Benchi and Cerami [1] and Zhu Xi-Ping and Cao Daomin [1]. See also Ding and Ni [1].

14.1 Setting up the Problem. An Additional Problem. The Basic Theorem

We shall demonstrate in detail the concentration compactness method on the following example:

$$-\Delta u + a(\mathbf{x})|u|^{p-2}u - b(\mathbf{x})|u|^{q-2}u = 0 \qquad \text{in } \mathbb{R}^N, \tag{P}$$

where $N > 2$, $2 \leq p < q < 2^*$, and $a, b : \mathbb{R}^N \to \mathbb{R}$ are measurable locally bounded nonnegative functions.

We impose on a and b the conditions analogous to those of §11 but simplified: we assume there are positive constants $\bar{a}$ and $\bar{b}$ such that

(*a*)

$$\lim_{R\to\infty} \sup_{\mathbf{x}\in\mathbb{R}^N\setminus B_R} |a(\mathbf{x}) - \bar{a}| = 0,$$

(*b*)

$$\lim_{R\to\infty} \sup_{\mathbf{x}\in\mathbb{R}^N\setminus B_R} |b(\mathbf{x}) - \bar{b}| = 0.$$

Consider also the additional problem

$$\begin{aligned} &- \Delta u + \bar{a}\,|u|^{p-2}u - \bar{b}\,|u|^{q-2}u = 0 \qquad \text{in } \mathbb{R}^N, \\ &u \to 0 \qquad \text{as } |\mathbf{x}| \to \infty, \end{aligned} \tag{P$^\infty$}$$

with $\bar{a}$ and $\bar{b}$ just defined.

The problem functionals

$$\Phi(u) = \frac{1}{2}\|\nabla u\|_2^2 + \frac{1}{p}\|a^{1/p}u\|_p^p - \frac{1}{q}\|b^{1/q}u\|_q^q,$$

$$\bar{\Phi}(u) = \frac{1}{2}\|\nabla u\|_2^2 + \frac{1}{p}\|\bar{a}^{1/p}u\|_p^p - \frac{1}{q}\|\bar{b}^{1/q}u\|_q^q$$

are associated with problems (P) and (P$^\infty$), respectively.

Denote

$$I_0(\bar{\Phi}) = \begin{cases} \inf\limits_{u\in\mathcal{P}(\bar{\Phi})} \bar{\Phi}, & \mathcal{P}(\bar{\Phi}) \neq \emptyset, \\ +\infty, & \mathcal{P}(\bar{\Phi}) = \emptyset, \end{cases}$$

where $\mathcal{P}(\bar{\Phi})$ is the set of all nontrivial critical points $u \in \mathcal{E}_{p,q}$ of the functional $\bar{\Phi} : \mathcal{E}_{p,q} \to \mathbb{R}$.

The behavior of PS-sequences is described by the following theorem, which is due to Benchi and Cerami [1] and Zhu Xi-Ping and Cao Daomin [1]:

THEOREM 14.1. *Let assumptions (a) and (b) hold. Let $I_0(\bar{\Phi}) > 0$. Let $\{u_n\} \subset \mathcal{E}_{p,q}$ be a PS_c-sequence of the functional Φ.*

Then there are a number $k \in \mathbb{N} \cup \{0\}$ and k sequences $\{\mathbf{y}_n^i\}$, $i = 1, \dots, k$, from $\mathbb{R}^N$ with $|\mathbf{y}_n^i| \to \infty$ and $|\mathbf{y}_n^i - \mathbf{y}_n^j| \to \infty$ for $j \neq i$ as $n \to \infty$; and there are k nontrivial solutions $u_{(j)} \in \mathcal{E}_{p,q}$ to problem (P^∞) and a solution $u_{(0)} \in \mathcal{E}_{p,q}$ to problem (P) such that for a subsequence $\{u_n\}$ the following is true:

(I) $\lim_{n\to\infty} \|u_n - u_{(0)} - \sum_{j=1}^k u_{(j)}(\,.\, - \mathbf{y}_n^j)\|_{\mathcal{E}_{p,q}} = 0,$

(II) $\Phi(u_{(0)}) + \sum_{j=1}^k \bar{\Phi}(u_{(j)}) = c.$

Remark. In §15 we shall show that $I_0(\Phi) > 0$ is always true for (P^∞) with $2 \le p < q < 2^*$. Hence, this condition actually may be omitted.

The following result, which is a direct consequence of Theorem 14.1, is also of interest.

14.2 AN ALTERNATIVE FORM OF THE THEOREM ON THE BEHAVIOR OF PS-SEQUENCES

THEOREM 14.2. *Let the assumptions of Theorem 14.1 be true. Then there are a number $k \in \mathbb{N} \cup \{0\}$ and k sequences $\{w_{n(i)}\} \subset \mathcal{E}_{p,q}$ and k sequences $\{\mathbf{y}_n^i\}$, $i = 1, \dots, k$, from $\mathbb{R}^N$, with $|\mathbf{y}_n^i| \to \infty$ and $|\mathbf{y}_n^i - \mathbf{y}_n^j| \to \infty$ for $j \neq i$ as $n \to \infty$, such that*

$$u_n = \sum_{i=0}^k w_{n(i)}(\,.\, - \mathbf{y}_n^i) \tag{14.2}$$

and

$$\begin{aligned} &w_{n(i)} \to u_{(i)} \quad \text{in } \mathcal{E}_{p,q}, \quad i = 1, \dots, k, \\ &w_{n(0)} \to u_{(0)} \quad \text{in } \mathcal{E}_{p,q}, \end{aligned} \tag{14.3}$$

where $u_{(0)}$ is a solution to problem (P), and $u_{(i)}$ are nontrivial solutions to problem (P^∞).

Proof. We shall first show how Theorem 14.2 follows from the basic Theorem 14.1. Indeed, denote $\mathbf{y}_n^0 = 0$ and define $w_{n(i)}$ as

$$w_{n(i)} = u_n(\,.\, + \mathbf{y}_n^i) - \sum_{j=0, j\neq i}^k u_{(j)}(\,.\, + \mathbf{y}_n^i - \mathbf{y}_n^j). \tag{14.4}$$

By virtue of (I) and (14.4),

$$w_{n(i)} \to u_{(i)} \quad \text{in } \mathcal{E}_{p,q}, \quad i = 1, \dots, k, \tag{14.5}$$

as $n \to \infty$ and analogously

$$w_{n(0)} \to u_{(0)} \quad \text{in } \mathcal{E}_{p,q}. \tag{14.6}$$

Now dwell upon (14.2). We have

$$\lim_{n\to\infty}\left(u_n-\sum_{i=0}^{k} w_{n(i)}(\cdot-\mathbf{y}_n^i)\right)$$

$$\overset{(14.4)}{=}\lim_{n\to\infty}\left(u_n-\sum_{i=0}^{k} u_n(\cdot+\mathbf{y}_n^i-\mathbf{y}_n^i)+\sum_{i=0}^{k}\Big(\sum_{j=0,j\neq i}^{k} u_{(j)}(\cdot-\mathbf{y}_n^j+\mathbf{y}_n^i-\mathbf{y}_n^i)\Big)\right)$$

$$\overset{(I)}{=}\lim_{n\to\infty}\left(u_n-(k+1)u_n+\sum_{i=0}^{k}(u_n-u_{(i)}(\cdot-\mathbf{y}_n^i))\right)$$

$$\overset{(I)}{=}\lim_{n\to\infty}(u_n-(k+1)u_n+(k+1)u_n-u_n)=0.$$

Hence,

$$u_n=\sum_{i=0}^{k} w_{n(i)}(\cdot-\mathbf{y}_n^i)+\bar{w}_n, \tag{14.7}$$

where $\bar{w}_n\to 0$ in $\mathcal{E}_{p,q}$. Redenote $w_{n(0)}+\bar{w}_n$ as $w_{n(0)}$. Then (14.5) and (14.6) turn into (14.3), and (14.7) becomes (14.2). □

Now we start the proof of Theorem 14.1. Preliminarily we shall formulate and prove some lemmas.

14.3 The First Lemma

LEMMA 14.3. *Let the assumptions of Theorem 14.1 be satisfied. Let $\{u_n\}$ be a PS-sequence of the functional Φ. Then it is possible to extract from $\{u_n\}$ a subsequence such that*

(i) $u_n\rightharpoonup u$ *in* $\mathcal{E}_{p,q}$,

(ii) for any ball $B_R\subset\mathbb{R}^N$ of the radius R centered at the origin of $\mathbb{R}^N$,

$$u_n\to u \qquad \text{in } \mathcal{E}_{p,q}(B_R),$$

(iii) $\Phi'(u)=0$ *in* $\mathcal{E}^*_{p,q}$.

Remark. Lemma 14.3 gives us a bit more information than we need to prove Theorem 14.1. Namely, statement *(ii)* is not necessary for this purpose. Note also that statement *(iii)* follows easily from *(i)* without reference to *(ii)* in a manner analogous to that of Section 3.4.

We present here the full proof of Lemma 14.3 as it may be helpful in generalizing to the case where, for example, instead of $-\Delta u$ the nonlinear operator $-\Delta_p u$ appears.

Proof of Lemma 14.3. First, we shall prove *(i)*. Estimating

$$\Phi(u_n)-\theta\,\langle\Phi'(u_n),u_n\rangle$$

for arbitrary $1/q < \theta < 1/p$ we find by virtue of the definition of a PS-sequence that

$$\begin{aligned}
&\left(\frac{1}{2}-\theta\right)\|\nabla u_n\|_2^2+\left(\frac{1}{p}-\theta\right)\|a^{1/p}u_n\|_p^p \\
&\quad\le\left(\frac{1}{2}-\theta\right)\|\nabla u_n\|_2^2+\left(\frac{1}{p}-\theta\right)\|a^{1/p}u_n\|_p^p-\left(\frac{1}{q}-\theta\right)\|b^{1/q}u_n\|_q^q \\
&\quad\le \mathbf{C}+\mu_n+\langle\xi_n,u_n\rangle\le\mathbf{C}+\mu_n+\|u_n\|_{\mathcal{E}_{p,q}}\|\xi_n\|_{\mathcal{E}^*_{p,q}},
\end{aligned} \tag{14.8}$$

because $\mu_n \to 0$ in $\mathbb{R}$ and $\xi_n \to 0$ in $\mathcal{E}^*_{p,q}$.

From (14.8) and Sobolev embedding the sequence $\{u_n\}$ is bounded in the norm of $\mathcal{E}_{p,q}$:

$$\|u_n\|_{\mathcal{E}_{p,q}} \le \mathbf{C}_0\,. \tag{14.9}$$

Because the space $\mathcal{E}_{p,q}$ is reflexive, one can choose a subsequence satisfying (i).

Now dwell on (ii). Let $\{u_n\}$ be a PS-sequence for Φ in $\mathcal{E}_{p,q}$.

Fix arbitrary $R > 1$. Choose arbitrary $1 > \varepsilon > 0$. We shall find such a n_0 that for all $m \ge n \ge n_0$

$$\|u_n - u_m\|_{\mathcal{E}_{p,q}}(B_R) \le \varepsilon. \tag{14.10}$$

For this purpose consider a set of smooth functions $\varphi_i : \mathbb{R}^N \to \mathbb{R}$ with $\varphi_i \equiv 0$ for $|\mathbf{x}| \ge R(i+1)$, $\varphi_i \equiv 1$ for $|\mathbf{x}| \le Ri$, and $0 \le \varphi_i \le 1$, $|\nabla\varphi_i| \le 2$, for $Ri \le |\mathbf{x}| \le R(i+1)$, where

$$i = 1,2,\dots,i_0 \equiv \left[2\frac{\mathbf{C}_0}{\varepsilon}\right]+1, \tag{14.11}$$

where $\mathbf{C}_0$ is as in (14.9) and $[\,.\,]$ is the integer part.

From the definition of PS-sequences we have

$$\begin{aligned}
&\|\xi_n-\xi_m\|_{\mathcal{E}^*_{p,q}}\|(u_n-u_m)\varphi_i\|_{\mathcal{E}_{p,q}} \ge |\langle\Phi'(u_n)-\Phi'(u_m),(u_n-u_m)\varphi_i\rangle| \\
&\quad=\Bigg|\|\nabla(u_n-u_m)\|_2^2(B_{Ri})+\int\limits_{B_{R(i+1)}\setminus B_{Ri}}(\nabla(u_n-u_m),\nabla((u_n-u_m)\varphi_i))\,d\mathbf{x} \\
&\quad+\int\limits_{B_{R(i+1)}} a(\mathbf{x})\,(|u_n|^{p-2}u_n-|u_m|^{p-2}u_m)(u_n-u_m)\varphi_i\,d\mathbf{x} \\
&\quad-\int\limits_{B_{R(i+1)}} b(\mathbf{x})\,(|u_n|^{q-2}u_n-|u_m|^{q-2}u_m)(u_n-u_m)\varphi_i\,d\mathbf{x}\Bigg|,
\end{aligned} \tag{14.12}$$

where

$$\|\xi_n-\xi_m\|_{\mathcal{E}^*_{p,q}} \to 0 \qquad \text{as } m \ge n \ge n_0 \to \infty. \tag{14.13}$$

Because $2 \leq p < q < 2^*$ and a is locally bounded, Sobolev embedding and the properties of norms yield

$$\begin{aligned}\|(u_n - u_m)\varphi_i\|_{\mathcal{E}_{p,q}} &\leq \mathbf{C}\|\nabla((u_n - u_m)\,\varphi_i)\|_2(B_{R(i+1)})\\ &\leq \mathbf{C}(\|\nabla(u_n - u_m)\varphi_i\|_2(B_{R(i+1)}) + \|(u_n - u_m)\nabla\varphi_i\|_2(B_{R(i+1)} \setminus B_{Ri}))\end{aligned} \tag{14.14}$$

Obviously, (14.14) implies

$$\|(u_n - u_m)\varphi_i\|_{\mathcal{E}_{p,q}} \leq \mathbf{C}(\|\nabla(u_n - u_m)\|_2(B_{R(i+1)}) + 2\|u_n - u_m\|_2(B_{R(i+1)} \setminus B_{Ri})). \tag{14.15}$$

Because $i \leq i_0$, the Sobolev embedding theorem for bounded domains and (14.9) provide the boundedness of (14.15). Thus, relations (14.13) and (14.9) imply that for sufficiently large n_0,

$$\|\xi_n - \xi_m\|_{\mathcal{E}^*_{p,q}}\|(u_n - u_m)\varphi_i\|_{\mathcal{E}_{p,q}} \leq \varepsilon. \tag{14.16}$$

Due to the Kondrashov theorem it follows from (i) that for i_0 defined in (14.11)

$$u_n \to u \qquad \text{in } L_q(B_{R(i_0+1)}). \tag{14.17}$$

Using the Hölder inequality and local boundedness of a and b we have

$$\left|\int\limits_{B_{R(i+1)}} b(\mathbf{x})\,(|u_n|^{q-2}u_n - |u_m|^{q-2}u_m)(u_n - u_m)\varphi_i\,d\mathbf{x}\right|$$
$$\leq \mathbf{C}\left(1 + \|b^{1/q}u_n\|_q^{q-1} + \|b^{1/q}u_m\|_q^{q-1}\right)\|u_n - u_m\|_q(B_{R\,(i_0+1)}),$$

$$\left|\int\limits_{B_{R(i+1)}} a(\mathbf{x})\,(|u_n|^{p-2}u_n - |u_m|^{p-2}u_m)\,(u_n - u_m)\varphi_i\,d\mathbf{x}\right|$$
$$\leq \mathbf{C}\left(1 + \|a^{1/p}u_n\|_p^{p-1} + \|a^{1/p}u_m\|_p^{p-1}\right)\|u_n - u_m\|_p(B_{R\,(i_0+1)}), \tag{14.18}$$

where $\mathbf{C}$ does not depend on u_n, u_m, and i. By (14.9) and (14.17), these terms can be estimated from above by ε for sufficiently large n_0.

Analogously to (14.15), we get for all $i \leq i_0$

$$\left|\int\limits_{B_{R(i+1)}\setminus B_{Ri}} (\nabla(u_n - u_m), \nabla((u_n - u_m)\varphi_i))\,d\mathbf{x}\right|$$
$$\leq 2\|\nabla(u_n - u_m)\|_2(B_{R(i+1)} \setminus B_{Ri})\,\|u_n - u_m\|_2(B_{R(i+1)})$$
$$+ \|\nabla(u_n - u_m)\|_2^2\,(B_{R(i+1)} \setminus B_{Ri}). \tag{14.19}$$

Relation (14.17) implies that for sufficiently large n_0 the first term on the right-hand side of (14.19) is less than ε. Now we estimate the second term on the right-hand side of (14.19). By virtue of (14.9) and (14.11) for any m, n there exists a number i such that

$$\|\nabla(u_n - u_m)\|_2(B_{R_{i+1}} \setminus B_{R_i}) \leq \frac{2\mathbf{C}_0}{i_0} \leq \varepsilon. \tag{14.20}$$

Gather these estimates. Given $1 > \varepsilon > 0$, we may choose i_0 as in (14.11) such that $2\mathbf{C}_0/i_0 \leq \varepsilon$, where $\mathbf{C}_0$ does not depend on ε and i_0. Then we may choose i such that (14.20) holds. Then we may choose n_0 such that (14.16) is valid and the right-hand sides of (14.18) and (14.19) can be estimated from above by ε for $m \geq n \geq n_0$. Hence, for any $1 > \varepsilon > 0$ and $R > 1$ there is a number n_0 such that for $m \geq n \geq n_0$,

$$\|\nabla(u_n - u_m)\|_2(B_R) \leq \mathbf{C}\varepsilon.$$

By embedding, this implies

$$\|\nabla(u_n - u_m)\|_2(B_R) + \|a^{1/p}(u_n - u_m)\|_p(B_R) \leq \mathbf{C}\varepsilon,$$

with $\mathbf{C}$ not depending on u_n, u_m, and ε. Hence, (14.10) is proved.

By force of (14.10) the sequence $\{u_n\}$ is a Cauchy sequence in the space $\mathcal{E}_{p,q}(B_R)$, that is, *(ii)* holds.

In order to prove *(iii)* it is sufficient to show, by virtue of the density of $\mathcal{D}(\mathbb{R}^N)$ in $\mathcal{E}_{p,q}$, that for any $\varphi \in \mathcal{D}(\mathbb{R}^N)$,

$$\langle \Phi'(u), \varphi \rangle = 0. \tag{14.21}$$

For any $\varphi \in \mathcal{D}(\mathbb{R}^N)$ there exists a radius R such that $\varphi \in \mathcal{D}(B_R)$, thus (14.21) follows from the definition of the PS-sequence because of the strong convergence *(ii)* and $\Phi \in C^1(\mathcal{E}_{p,q})$:

$$|\langle \Phi'(u), \varphi \rangle| \leq \lim_{n\to\infty} \|\Phi'(u_n)\|_{\mathcal{E}^*_{p,q}} \|\varphi\|_{\mathcal{E}_{p,q}} = 0. \qquad \square$$

Remark. Instead of the functional Φ the functional $\bar{\Phi}$ might be considered. Then the following lemma may be proved:

LEMMA 14.4. *Let the assumptions of Theorem 14.1 be satisfied. Let $\{u_n\}$ be a PS-sequence of the functional $\bar{\Phi}$ on $\mathcal{E}_{p,q}$. Then it is possible to extract from $\{u_n\}$ a subsequence such that*

(i) $u_n \rightharpoonup u$ *in* $\mathcal{E}_{p,q}$,

(ii) for any ball $B_R \subset \mathbb{R}^N$ of the radius R centered at the origin of $\mathbb{R}^N$

$$u_n \to u \qquad \text{in } \mathcal{E}_{p,q}(B_R),$$

(iii) $\bar{\Phi}'(u) = 0$ *in* $\mathcal{E}^*_{p,q}$.

14.4 THE SECOND LEMMA

The following result is a consequence of Lemma 14.4.

LEMMA 14.5. *For any PS-sequence* $\{u_n\}$ *of* Φ *in* $\mathcal{E}_{p,q}$ *there is* $u \in \mathcal{E}_{p,q}$ *and a subsequence of* $\{u_n\}$ *such that*

$$(i) \qquad \lim_{n\to\infty} (\Phi(u_n) - \Phi(u_n - u)) = \Phi(u),$$

$$(ii) \qquad \lim_{n\to\infty} (\Phi'(u_n) - \Phi'(u_n - u)) = \Phi'(u) \qquad \text{in } \mathcal{E}^*_{p,q}.$$

The analogous results hold for $\bar{\Phi}$.

Proof. We consider the case of Φ. Let $\{u_n\}$ be a PS-sequence. We first prove (i). From Lemma 14.3 it follows that

$$u_n \rightharpoonup u \qquad \text{in } \mathcal{E}_{p,q}.$$

Recall that the same situation was considered in Lemma 11.5 where the relation

$$\lim_{n\to\infty} (\|b^{1/q}u_n\|_q^q - \|b^{1/q}(u_n - u)\|_q^q) = \|b^{1/q}u\|_q^q \tag{14.22}$$

was proved. Analogously,

$$\lim_{n\to\infty} (\|a^{1/p}u_n\|_p^p - \|a^{1/p}(u_n - u)\|_p^p) = \|a^{1/p}u\|_p^p. \tag{14.23}$$

The first relation in (11.21) gives us also

$$\lim_{n\to\infty} (\|\nabla u_n\|_2^2 - \|\nabla(u_n - u)\|_2^2) = \|\nabla u\|_2^2. \tag{14.24}$$

Using (14.22), (14.23), and (14.24), we get (i).

Now we prove (ii) in a manner analogous to the proof of Lemma 11.5.

By Lemma 14.3,

$$\begin{aligned} &\|u_n\|_{\mathcal{E}_{p,q}} \le \mathbf{C}, \\ &u_n \rightharpoonup u \qquad \text{in } \mathcal{E}_{p,q}. \end{aligned} \tag{14.25}$$

Note that

$$\nabla u_n - \nabla(u_n - u) = \nabla u. \tag{14.26}$$

On the other hand, for any u_n and φ from $\mathcal{E}_{p,q}$,

$$\begin{aligned} &\int\limits_{\mathbb{R}^N\setminus B_R} b(\mathbf{x})|u_n|^{q-2}u_n\varphi\, d\mathbf{x} \\ &\qquad = \int\limits_{\mathbb{R}^N\setminus B_R} b(\mathbf{x})|u_n - u|^{q-2}(u_n - u)\varphi\, d\mathbf{x} \\ &\qquad\qquad + (q-1) \int\limits_{\mathbb{R}^N\setminus B_R} \int\limits_0^1 b(\mathbf{x})\, |u_n - u + \theta u|^{q-2}\, u\, \varphi\, d\theta\, d\mathbf{x}. \end{aligned}$$

Consequently,

$$\int_{\mathbb{R}^N} b(\mathbf{x})(|u_n|^{q-2}u_n - |u|^{q-2}u - |u_n - u|^{q-2}(u_n - u))\,\varphi\,d\mathbf{x}$$

$$= \int_{B_R} b(\mathbf{x})\,(|u_n|^{q-2}u_n - |u|^{q-2}u)\varphi\,d\mathbf{x}$$

$$- \int_{B_R} b(\mathbf{x})\,|u_n - u|^{q-2}(u_n - u)\varphi\,d\mathbf{x} - \int_{\mathbb{R}^N\setminus B_R} b(\mathbf{x})\,|u|^{q-2}u\varphi\,d\mathbf{x}$$

$$+ (q-1)\int_{\mathbb{R}^N\setminus B_R}\int_0^1 b(\mathbf{x})\,|u_n - u + \theta u|^{q-2}u\,\varphi\,d\theta\,d\mathbf{x}. \qquad (14.27)$$

Relation (14.25) implies

$$u_n \to u \qquad \text{in } L_q(B_R)$$

for any R. Hence, for any R the first and second integrals on the right-hand side of (14.27) because of the Vainberg-Krasnoselskii theorem may be estimated by

$$\mathbf{C}\,(\|b^{1/q}\varphi\|_q\|b^{\frac{q-1}{q}}(|u_n|^{q-2}u_n - |u|^{q-2}u)\|_{q/(q-1)}(B_R) + \|b^{1/q}\varphi\|_q\,\|b^{1/q}(u_n - u)\|_q^{q-1}(B_R)) \le \mathbf{C}\|\varphi\|_q\,\varepsilon_n,$$

where $\mathbf{C}$ does not depend on u, and ε_n tend to zero as $n \to \infty$. The third integral tends to zero as $R \to \infty$ because of the definition of $L_{q(b)}(\mathbb{R}^N)$. In order to estimate the last integral, we make use of the Hölder inequality and boundedness, by (14.25), of $\{u_n\}$ in $L_q(\mathbb{R}^N)$. We get

$$\left|\int_{\mathbb{R}^N\setminus B_R}\int_0^1 b(\mathbf{x})\,|u_n - u + \theta u|^{q-2}u\,\varphi\,d\mathbf{x}\,d\theta\right|$$

$$\le \max_{0\le\theta\le 1}\left(\|b^{1/q}(u_n - u + \theta u)\|_q^{q-2}(\mathbb{R}^N\setminus B_R)\right)\|b^{1/q}u\|_q(\mathbb{R}^N\setminus B_R)\,\|b^{1/q}\varphi\|_q$$

$$\le \mathbf{C}\,(\|b^{1/q}u_n\|_q^{q-2} + \|b^{1/q}u\|_q^{q-2})\,\|b^{1/q}u\|_q(\mathbb{R}^N\setminus B_R)\,\|b^{1/q}\varphi\|_q$$

$$\le \mathbf{C}\,\|b^{1/q}u\|_q(\mathbb{R}^N\setminus B_R)\,\|b^{1/q}\varphi\|_q = \mathbf{C}(R)\|b^{1/q}\varphi\|_q,$$

where $\mathbf{C}(R) \to 0$ as $R \to \infty$ uniformly in n.

Thus, choosing an arbitrary $\varepsilon > 0$, we may find R_ε such that for all n the absolute values of the third and fourth integrals on the right-hand side of (14.27) are less than $\varepsilon\,\|\varphi\|_q$. Given ε and R_ε, we find a number n_ε such that, for all $n \ge n_\varepsilon$, the absolute values of the first and second integrals are also less than $\varepsilon \|\varphi\|_q$. After taking the limit $\varepsilon \to 0$, we get

$$\lim_{n\to\infty} \sup_{\|\varphi\|_q=1} \int_{\mathbb{R}^N} b(\mathbf{x})(|u_n|^{q-2}u_n - |u_n - u|^{q-2}(u_n - u) - |u|^{q-2}u)\,\varphi\,d\mathbf{x} = 0. \quad (14.28)$$

Analogously,

$$\lim_{n\to\infty} \sup_{\|\varphi\|_p=1} \int_{\mathbb{R}^N} a(\mathbf{x})(|u_n|^{p-2}u_n - |u_n - u|^{p-2}(u_n - u) - |u|^{p-2}u)\,\varphi\,d\mathbf{x} = 0. \quad (14.29)$$

Taking into consideration (14.26), (14.28), and (14.29), we get (ii). □

14.5 The Third Lemma

We prove another preliminary result, analogous to Lemma 11.6.

LEMMA 14.6. *Let a sequence* $\{u_n\} \subset \mathcal{E}_{p,q}$ *satisfy*

$$u_n \rightharpoonup 0 \qquad \text{in } \mathcal{E}_{p,q}. \quad (14.30)$$

Then

$$(i) \qquad \lim_{n\to\infty} |\Phi(u_n) - \bar{\Phi}(u_n)| = 0,$$

$$(ii) \qquad \lim_{n\to\infty} \|\Phi'(u_n) - \bar{\Phi}'(u_n)\|_{\mathcal{E}^*_{p,q}} = 0,$$

where Φ *and* $\bar{\Phi}$ *are the same as in Theorem 14.1.*

Proof. It is easy to see that (i) is a corollary of relations similar to (11.15), which may be proved as in Lemma 11.6.

Relation (ii) may be proved in the same manner. By (a) of Section 14.1, for any $\varepsilon > 0$ there exists R_ε such that

$$\sup_{\mathbf{x}\in\mathbb{R}^N\setminus B_{R_\varepsilon}} |a(\mathbf{x}) - \bar{a}| \le \varepsilon,$$

that is, for this radius, all u_n, and $\varphi \in \mathcal{E}_{p,q}$,

$$\left| \int_{\mathbb{R}^N\setminus B_R} (a(\mathbf{x}) - \bar{a})\,|u_n|^{p-2}u_n\varphi\,d\mathbf{x} \right|$$

$$\le \varepsilon \int_{\mathbb{R}^N\setminus B_R} |u_n|^{p-1}|\varphi|\,d\mathbf{x} \le \varepsilon\,\|u_n\|_p^{p-1}\|\varphi\|_p. \quad (14.31)$$

On the other hand, by the Kondrashov theorem, (14.30) implies

$$u_n \to u \qquad \text{in } L_q(B_R) \tag{14.32}$$

for any B_R. At last, by (14.30),

$$\|a^{1/p} u_n\|_p \le \mathbf{C}.$$

Now, given ε, we may by (14.31) choose R_ε, and then by (14.32) choose a number n_ε such that for all $n \ge n_\varepsilon$ the estimate

$$\begin{aligned}
\left| \int\limits_{\mathbb{R}^N} a(\mathbf{x})\, |u_n|^{p-2} u_n - \bar{a}(|\mathbf{x}|)\, |u_n|^{p-2} u_n\, \varphi\, d\mathbf{x} \right| \\
&\le \int\limits_{\mathbb{R}^N \setminus B_{R_\varepsilon}} |a(\mathbf{x}) - \bar{a}(|\mathbf{x}|)|\, |u_n|^{p-1} |\varphi|\, d\mathbf{x} \\
&\quad + \int\limits_{B_{R_\varepsilon}} |a(\mathbf{x}) - \bar{a}(|\mathbf{x}|)|\, |u_n|^{p-1} |\varphi|\, d\mathbf{x} \\
&\le (\varepsilon\, \|u_n\|_p^{p-1} + \varepsilon) \|\varphi\|_p \le \varepsilon \mathbf{C}\, \|\varphi\|_p
\end{aligned}$$

holds, where $\mathbf{C}$ does not depend on n, ε, and φ. Estimating analogously the term with b and taking the limit $\varepsilon \to 0$, we get the result. □

14.6 THE LIONS LEMMA

Finally, in this section one of the basic lemmas for the concentration compactness method will be introduced. This lemma denies a sequence $\{u_n\} \subset \mathcal{E}_{p,q}$ to distribute itself over $\mathbb{R}^N$.

LIONS LEMMA. *Let $1 \le p < 2^*$. Suppose that for all $u_n \in \mathcal{E}_{p,q}$ the estimates*

$$\|\nabla u_n\|_2 \le \mathbf{C}, \qquad \|u_n\|_p \le \mathbf{C} \tag{14.33}$$

hold. Let for some $R > 0$ the relation

$$\sup_{\mathbf{y} \in \mathbb{R}^N} \int\limits_{\mathbf{y} + B_R} (|\nabla u_n|^2 + |u_n|^p)\, d\mathbf{x} \to 0 \qquad \text{as } n \to \infty \tag{14.34}$$

be valid. Then for any $p < q < 2^*$,

$$u_n \to 0 \qquad \text{in } L_q(\mathbb{R}^N) \qquad \text{as } n \to \infty. \tag{14.35}$$

Proof. We cover $\mathbb{R}^N$ by balls B_R^i in such a way that every point $\mathbf{x} \in \mathbb{R}^N$ should be contained in no more then K balls. By the Sobolev embedding theorem, for any R there is a constant $\mathbf{C}$ such that for all $u \in \mathcal{E}_{p,q}$ and i, the inequality

$$\|u\|_q^q(B_R^i) \leq \mathbf{C}\,(\|\nabla u\|_2^q + \|u\|_p^q)\,(B_R^i) \tag{14.36}$$

holds. By (14.36), we have

$$\begin{aligned}\|u_n\|_q^q(\mathbb{R}^N) &\leq \sum_i \|u_n\|_q^q(B_R^i)\\ &\leq \mathbf{C}\sum_i \left(\|\nabla u_n\|_2^2\,\|\nabla u_n\|_2^{q-2} + \|u_n\|_p^p\,\|u_n\|_p^{q-p}\right)(B_R^i).\end{aligned}$$

Hence, by (14.33) and (14.34),

$$\begin{aligned}\|u_n\|_q^q(\mathbb{R}^N) &\leq \mathbf{C}\,\big(\sup_i \|\nabla u_n\|_2^{q-2}(B_R^i)\,\|\nabla u_n\|_2^2(\mathbb{R}^N)\\ &\quad + \sup_i \|u_n\|_p^{q-p}(B_R^i)\,\|u_n\|_p^p(\mathbb{R}^N)\big)\\ &\leq \mathbf{C}\sup_i \left(\|\nabla u_n\|_2^{q-2} + \|u_n\|_p^{q-p}\right)(B_R^i) \to 0 \qquad \text{as } n \to \infty. \qquad \square\end{aligned}$$

Remark. An analogous result is true for $2^* < q < p < +\infty$. These results are due to Lions [1].

Remark. If $p = q$, (14.35) is not generally true. Is is sufficient to consider a sequence

$$u_n(\mathbf{x}) = \frac{\varphi_n(\mathbf{x})}{1 + |\mathbf{x}|^{N/2}}(\ln(1+n))^{-1/2},$$

where φ_n are defined as

$$\varphi_n(\mathbf{x}) = \begin{cases} 1, & |\mathbf{x}| \leq n,\\ n+1-|\mathbf{x}|, & n \leq |\mathbf{x}| \leq n+1,\\ 0, & n+1 \leq |\mathbf{x}|.\end{cases}$$

For this sequence, (14.33) and (14.34) hold for $p = 2$ but (14.35) with $p = q$ is not valid.

14.7 PROOF OF THEOREM 14.1

Now we can start the proof of Theorem 14.1.

Let $\{u_n\}$ be a PS$_c$-sequence of the problem functional for (P) on $\mathcal{E}_{p,q}$. By virtue of Lemma 14.3, statements (*i*) and (*iii*), the following is true up to a subsequence

$$u_n \rightharpoonup u_{(0)} \qquad \text{in } \mathcal{E}_{p,q}, \tag{14.37}$$

where $u_{(0)}$ is a solution of (P).

Letting

$$v_n^1 = u_n - u_{(0)} \tag{14.38}$$

we get from Lemma 14.5, (14.37), and the definition of PS-sequences,

$$\lim_{n\to\infty} \Phi(v_n^1) = c - \Phi(u_{(0)}),$$
$$\lim_{n\to\infty} \Phi'(v_n^1) = 0 - \Phi'(u_{(0)}) \qquad \text{in } \mathcal{E}_{p,q}^*.$$

Denoting

$$c_1 = c - \Phi(u_{(0)}) \tag{14.39}$$

and taking into account that $\Phi'(u_{(0)}) = 0$, we have

$$\begin{aligned} \lim_{n\to\infty} \Phi(v_n^1) &= c_1 \qquad \text{in } \mathbb{R}, \\ \lim_{n\to\infty} \Phi'(v_n^1) &= 0 \qquad \text{in } \mathcal{E}_{p,q}^*. \end{aligned} \tag{14.40}$$

Then, in accordance with (14.37) and (14.38),

$$v_n^1 \rightharpoonup 0 \qquad \text{in } \mathcal{E}_{p,q},$$

which implies the correctness of the assumptions of Lemma 14.6, and consequently, by virtue of (14.40),

$$\begin{aligned} \lim_{n\to\infty} \bar{\Phi}(v_n^1) &= c_1 \qquad \text{in } \mathbb{R}, \\ \lim_{n\to\infty} \bar{\Phi}'(v_n^1) &= 0 \qquad \text{in } \mathcal{E}_{p,q}^*. \end{aligned} \tag{14.41}$$

For $\{v_n^1\}$ two variants are possible:

(*i*) for all R $\displaystyle\lim_{n\to\infty} \sup_{\mathbf{y}\in\mathbb{R}^N} \int_{\mathbf{y}+B_R} (|\nabla v_n^1|^2 + |v_n^1|^p)\, d\mathbf{x} = 0,$

(*ii*) there is an R such that $\displaystyle\overline{\lim_{n\to\infty}} \sup_{\mathbf{y}\in\mathbb{R}^N} \int_{\mathbf{y}+B_R} (|\nabla v_n^1|^2 + |v_n^1|^p)\, d\mathbf{x} > 0.$

In the first case, in accordance with the Lions lemma and the boundedness of the sequence $\{v_n^1\}$ in $\mathcal{E}_{p,q}$,

$$\|v_n^1\|_q \to 0 \qquad \text{as } n \to \infty,$$

holds and, consequently, by the Hölder inequality,

$$\left| \int_{\mathbb{R}^N} \bar{b}\, (|v_n^1|^{q-2} v_n^1 - |v_m^1|^{q-2} v_m^1)(v_n^1 - v_m^1)\, d\mathbf{x} \right|$$
$$\le \mathbf{C}(\|v_n^1\|_q^{q-1} + \|v_m^1\|_q^{q-1})\, \|v_n^1 - v_m^1\|_q \to 0 \qquad \text{as } m \ge n \to \infty.$$

Hence, we have, by (14.41) and the boundedness of $\{v_n^1\}$ in $\mathcal{E}_{p,q}$,

$$\begin{aligned}
&\|\nabla(v_n^1 - v_m^1)\|_2^2 + \mathbf{C}\,\|v_n^1 - v_m^1\|_p^p \\
&\quad \le \|\nabla(v_n^1 - v_m^1)\|_2^2 + \int\limits_{\mathbb{R}^N} \bar{a}(|v_n^1|^{p-2}v_n^1 - |v_m^1|^{p-2}v_m^1)(v_n^1 - v_m^1)\,d\mathbf{x} \\
&\quad = \langle \Phi'(v_n^1) - \Phi'(v_m^1), v_n^1 - v_m^1 \rangle \\
&\qquad + \int\limits_{\mathbb{R}^N} \bar{b}\,(|v_n^1|^{q-2}v_n^1 - |v_m^1|^{q-2}v_m^1)(v_n^1 - v_m^1)\,d\mathbf{x} \to 0 \qquad \text{as } m \ge n \to \infty,
\end{aligned}$$

i.e., $\{v_n^1\}$ is a Cauchy sequence and

$$v_n^1 \to 0 \qquad \text{in } \mathcal{E}_{p,q}. \tag{14.42}$$

By force of (14.38) statement (II) of Theorem 14.1 then holds for $k = 0$, and the proof is thus complete.

Let now (*ii*) be true. Then there is a sequence $\{\mathbf{y}_n^1\} \subset \mathbb{R}^N$ such that for a subsequence of $\{v_n^1\}$,

$$\lim_{n\to\infty} \|u_{n(1)}\|_{\mathcal{E}_{p,q}}(B_R) > 0, \tag{14.43}$$

where

$$u_{n(1)}(\mathbf{x}) = v_n^1(\mathbf{x} + \mathbf{y}_n^1).$$

(We have $|\mathbf{y}_n^1| \to \infty$ as $n \to \infty$, otherwise (14.43) contradicts $v_n^1 \rightharpoonup 0$ in $\mathcal{E}_{p,q}$.)

Then by force of the invariance of $\bar{\Phi}$ with respect to translations we have analogously to (14.41)

$$\begin{aligned}
&\lim_{n\to\infty} \bar{\Phi}(u_{n(1)}) = c_1 \qquad \text{in } \mathbb{R}, \\
&\lim_{n\to\infty} \bar{\Phi}'(u_{n(1)}) = 0 \qquad \text{in } \mathcal{E}_{p,q}^*.
\end{aligned} \tag{14.44}$$

By force of invariance of the norm with respect to translations we have also

$$\|u_{n(1)}\|_{\mathcal{E}_{p,q}} \le \mathbf{C}\,. \tag{14.45}$$

By virtue of (14.45) we may choose a subsequence $\{u_{n(1)}\}$ such that

$$u_{n(1)} \rightharpoonup u_{(1)} \qquad \text{in } \mathcal{E}_{p,q}.$$

From (14.44) $\{u_{n(1)}\}$ is a PS_{c_1}-sequence for $\bar{\Phi}$ in $\mathcal{E}_{p,q}$, and then Lemma 14.4 implies that

$$\bar{\Phi}'(u_{(1)}) = 0 \qquad \text{in } \mathcal{E}_{p,q}^*,$$

where $u_{(1)} \not\equiv 0$ from (14.43).

Denoting

$$v_n^2 = u_{n(1)} - u_{(1)},$$

we repeat the procedure, getting

$$\begin{aligned} \lim_{n\to\infty} \bar{\Phi}(v_n^2) &= c_1 - \bar{\Phi}(u_{(1)}) && \text{in } \mathbb{R}, \\ \lim_{n\to\infty} \bar{\Phi}'(v_n^2) &= 0 - \bar{\Phi}'(u_{(1)}) && \text{in } \mathcal{E}_{p,q}^*, \end{aligned} \tag{14.46}$$

and then either

$$v_n^2 \to 0 \qquad \text{in } \mathcal{E}_{p,q} \qquad \text{as } n \to \infty$$

and thus $k = 1$ in the statement of Theorem 14.1, or for a subsequence and some R,

$$\lim_{n\to\infty} \sup_{\mathbf{y}\in\mathbb{R}^N} \int\limits_{\mathbf{y}+B_R} (|\nabla v_n^2|^2 + |v_n^2|^p)\, d\mathbf{x} > 0.$$

In the second case we build analogously to $\{u_{n(1)}\}$ a sequence $\{u_{n(2)}\}$ and so on. By force of (14.39), (14.41), (14.44), (14.46), and analogous relations for v_n^k with higher k, we have

$$c = \Phi(u_{(0)}) + \bar{\Phi}(u_{(1)}) + \bar{\Phi}(u_{(2)}) + \cdots$$

Because $\bar{\Phi}(u_{(j)}) \geq I_0(\bar{\Phi}) > 0$, and c is a finite number, this procedure has to be interrupted sooner or later, and we get (I) for some k. Finishing the procedure for given k means that variant (i) for $\{v_n^{k+1}\}$, where $v_n^{k+1} = u_{n(k)} - u_{(k)}$, holds in this case and hence

$$v_n^{k+1} \to 0 \qquad \text{in } \mathcal{E}_{p,q} \quad \text{as } n \to \infty.$$

Then (II) follows as well. □

14.8 A General Theorem

Theorems 14.1 and 14.2 may be generalized on the case of equations

$$-\Delta u + f(\mathbf{x}, u) = 0 \qquad \text{in } \mathbb{R}^N \tag{P}$$

with appropriate Carathéodory f.

Denote

$$F(\mathbf{x}, t) = \int\limits_0^t f(\mathbf{x}, s)\, ds.$$

(1) Let $f = f_1 + f_2$, where $f_1 \in K_{p,q}$, $f_2 \in K_q$ with $2 \leq p < q < 2^*$, $N > 2$, and $\partial f/\partial u \in K_{p-1,q-1}$.

(2) Let there be a constant $\theta < 1/2$, positive constants γ_1, γ_2, and a function $g_0 \in L_1(\mathbb{R}^N)$ such that

$$F(\mathbf{x},t) - \theta\, f(\mathbf{x},t)\, t \geq \gamma_1\, |t|^p + g_0(\mathbf{x}),$$
$$(f_1(\mathbf{x},t') - f_1(\mathbf{x},t''))(t'-t'') \geq \gamma_2\, |t'-t''|^p,$$

for almost all $\mathbf{x} \in \mathbb{R}^N$ and all $t, t', t'' \in \mathbb{R}$.

(3) Let there be a continuous function $\bar{f} : \mathbb{R} \to \mathbb{R}$ such that

$$\lim_{R\to\infty} \sup_{\substack{\mathbf{x}\in\mathbb{R}^N\setminus B_R,\\ t\in\mathbb{R}\setminus\{0\}}} \frac{|f(\mathbf{x},t) - \bar{f}(t)|}{|t|^{p-1} + |t|^{2^*-1}} = 0.$$

Consider the auxiliary problem

$$-\Delta u + \bar{f}(u) = 0 \qquad \text{in } \mathbb{R}^N \tag{P$^\infty$}$$

Define the problem functionals

$$\Phi(u) = \frac{1}{2}\|\nabla u\|_2^2 + \int_{\mathbb{R}^N} F(u)\, d\mathbf{x},$$

$$\bar{\Phi}(u) = \frac{1}{2}\|\nabla u\|_2^2 + \int_{\mathbb{R}^N} \bar{F}(u)\, d\mathbf{x},$$

where

$$\bar{F}(t) = \int_0^t \bar{f}(s)\, ds,$$

and define $I_0(\bar{\Phi})$ as

$$I_0(\bar{\Phi}) = \begin{cases} \inf\limits_{u\in\mathcal{P}(\bar{\Phi})} \bar{\Phi}, & \mathcal{P}(\bar{\Phi}) \neq \emptyset, \\ +\infty, & \mathcal{P}(\bar{\Phi}) = \emptyset, \end{cases}$$

where $\mathcal{P}(\bar{\Phi})$ is the set of all nontrivial critical points $u \in \mathcal{E}_{p,q}$ of the functional $\bar{\Phi} : \mathcal{E}_{p,q} \to \mathbb{R}$.

The following analog of Theorem 14.1 is true:

THEOREM 14.7. *Let all the assumptions of this section hold. Let $I_0(\bar{\Phi}) > 0$. Let $\{u_n\} \subset \mathcal{E}_{p,q}$ be a PS_c-sequence of the functional Φ.*

Then there are a number $k \in \mathbb{N} \cup \{0\}$ and k sequences $\{\mathbf{y}_n^i\}$, $i = 1, \ldots, k$, from $\mathbb{R}^N$ with $|\mathbf{y}_n^i| \to \infty$ and $|\mathbf{y}_n^i - \mathbf{y}_n^j| \to \infty$ for $j \neq i$ as $n \to \infty$; and there are k nontrivial solutions $u_{(j)} \in \mathcal{E}_{p,q}$ to problem (P^∞) and a solution $u_{(0)} \in \mathcal{E}_{p,q}$ to problem (P) such that, for a subsequence $\{u_n\}$, the following is true:

(I) $\lim_{n\to\infty} \|u_n - u_{(0)} - \sum_{j=1}^k u_{(j)}(\,\cdot\, - \mathbf{y}_n^j)\|_{\mathcal{E}_{p,q}} = 0,$

(II) $\Phi(u_{(0)}) + \sum_{j=1}^k \bar{\Phi}(u_{(j)}) = c.$

The theorem analogous to Theorem 14.2 is also true.

The proof repeats with appropriate modifications the proof of Theorems 14.1 and 14.2.

In the next two paragraphs we shall demonstrate the usefulness of these theorems.

§15. A General Comparison Theorem. The Ground State. Examples for the Mountain Pass Method

In this section we use the results of the previous one in order to formulate an abstract comparison theorem in the spirit of problems considered in §11. The notion of a *ground state* is introduced for an auxiliary variational problem and it is shown that this state exists for some problems. Then this theory is applied to concrete examples.

15.1 A COMPARISON THEOREM

Consider the problem

$$\begin{aligned} -\Delta u + f(\mathbf{x}, u) = 0 \qquad &\text{in } \mathbb{R}^N, \\ u \to 0 \qquad \text{as } |\mathbf{x}| \to \infty,& \end{aligned} \tag{P}$$

and the corresponding auxiliary one

$$\begin{aligned} -\Delta u + \bar{f}(u) = 0 \qquad &\text{in } \mathbb{R}^N, \\ u \to 0 \qquad \text{as } |\mathbf{x}| \to \infty.& \end{aligned} \tag{P$^\infty$}$$

All conditions on f and $\bar{f}$ are assumed to be the same as in Theorem 14.7. Recall that the functional Φ is associated with problem (P) and the functional $\bar{\Phi}$ is associated with problem (P$^\infty$); for $\bar{\Phi}$ the value $I_0(\bar{\Phi})$ is defined. Both functionals belong to $C^1(\mathcal{E}_{p,q})$.

We will say that Φ satisfies the PS$_{(a,b)}$-condition if it does so for all $c \in (a, b)$.

THEOREM 15.1. *Let the assumptions of Theorem 14.1 be true. Then if there is no solution $u \in \mathcal{E}_{p,q}$ to problem (P) with*

$$\Phi(u) < 0, \tag{15.1}$$

then the functional Φ satisfies the $PS_{(0,I_0(\bar{\Phi}))}$-condition on the space $\mathcal{E}_{p,q}$.

Proof. By the conditions of the theorem for any possible solution $u \in \mathcal{E}_{p,q}$ to (P),

$$\Phi(u) \geq 0. \tag{15.2}$$

Consider a PS$_c$-sequence $\{u_n\} \subset \mathcal{E}_{p,q}$ satisfying

$$\Phi(u_n) \to c \in (0, I_0(\bar{\Phi})) \qquad \text{in } \mathbb{R},$$
$$\Phi'(u_n) \to 0 \qquad \text{in } \mathcal{E}^*_{p,q}$$

as $n \to \infty$. By force of Theorem 14.7 there is its subsequence with

$$u_n - \sum_{j=1}^{k} u_{(j)}(\cdot - \mathbf{y}_n^j) \to u_{(0)} \qquad \text{in } \mathcal{E}_{p,q},$$

where $u_{(0)}$ is a solution of (P) and $u_{(j)}$ are the nontrivial solutions of (P$^\infty$). Hence, from relation (II) of §14, Theorem 14.7, and (15.2) it follows that

$$I_0(\bar{\Phi}) > c = \Phi(u_{(0)}) + \sum_{j=1}^{k} \bar{\Phi}(u_{(j)}) \geq \sum_{j=1}^{k} \bar{\Phi}(u_{(j)}) \geq k\, I_0(\bar{\Phi})$$

with $I_0(\bar{\Phi}) > 0$. Hence $k = 0$, that is, by force of (I) of §14

$$u_n \to u_{(0)} \qquad \text{in } \mathcal{E}_{p,q},$$

which means that the $PS_{(0,I_0(\bar{\Phi}))}$-condition is satisfied. □

15.2 A GROUND STATE

In order to endow $I_0(\bar{\Phi})$ with more concrete meaning we introduce the notion of a ground state. Dwell upon problem (P$^\infty$):

$$-\Delta u + \bar{f}(u) = 0 \qquad \text{in } \mathbb{R}^N, \tag{P$^\infty$}$$

where $f \in K_{pq}$ with $2 \leq p < q < 2^*$, and $N > 2$.

DEFINITION. A solution $u_0 \in \mathcal{E}_{p,q}$ is a *ground state* of problem (P$^\infty$) if for any other solution $u \in \mathcal{E}_{p,q}$ to (P$^\infty$) the inequalities

$$0 < \bar{\Phi}(u_0) \leq \bar{\Phi}(u) \tag{15.3}$$

hold.

Obviously, in accordance with the definition of Section 14.8, we have

$$I_0(\bar{\Phi}) = \bar{\Phi}(u_0).$$

The following result facilitates the implementation of Theorem 15.1. This observation is credited to Coleman, Glazer and Martin [1].

THEOREM 15.2. *Let f be a continuous odd function satisfying the assumptions of Theorem 9.10. Then problem (P^∞) has a ground state, $u_0 \in \mathcal{E}_{p,q}$, where p and q are defined in the assumptions of this theorem.*

Proof. In accordance with the approach of §9 the problem

$$\left\{u \in \mathcal{E}^{\mathrm{rad}}_{p,q}, \quad \min J_1(u), \quad J_2(u) = -1\right\} \tag{$V^{\infty,\mathrm{rad}}$}$$

with $J_1, J_2 : \mathcal{E}_{p,q} \to \mathbb{R}$,

$$J_1(u) = \|\nabla u\|_2^2,$$
$$J_2(u) = \int_{\mathbb{R}^N} \bar{F}(u)\, d\mathbf{x},$$

where

$$\bar{F}(t) = \int_0^t \bar{f}(s)\, ds$$

is associated with problem (P^∞). In §9 it was shown that this variational problem under the conditions of Theorem 9.10 has a solution $u_0 \in \mathcal{E}^{\mathrm{rad}}_{p,q}$, which is a nontrivial solution of the corresponding eigenvalue problem with some $\lambda > 0$. In §12 it was shown that the function v_0 defined for the eigenvalue $\lambda > 0$ as

$$v_0(\mathbf{x}) = u_0(\mathbf{x}/\sqrt{\lambda})$$

is a solution of the boundary problem (P^∞). We claim now that v_0 is a ground state.

Indeed, note that by force of Appendix D, Theorem D.4, any solution of variational problem ($V^{\infty,\mathrm{rad}}$) is also a solution of the variational problem

$$\{u \in \mathcal{E}_{p,q}, \quad \min J_1(u), \quad J_2(u) = -1\}. \tag{V^∞}$$

Let us study the properties of such solutions.

For this purpose define for any nontrivial solution $v \in \mathcal{E}_{p,q}$ of (P^∞) the function v_σ (cf. Section 9.2) as

$$v_\sigma(\mathbf{x}) = v(\mathbf{x}/\sigma),$$

where we assume

$$\sigma = (-J_2(v))^{-1/N}.$$

By scaling Lemma 9.2, we have

$$J_2(v_\sigma) = -1. \tag{15.4}$$

On the other hand, in §29 we shall show that any solution $v \in \mathcal{E}_{p,q}$ of (P^∞) satisfies the P-identity (29.7), which in our case takes the form

$$(2-N)\,J_1(v) = N\,J_2(v).$$

Scaling Lemma 9.2 and the P-identity provide

$$J_1(v_\sigma) = J_1(v)\,(-J_2(v))^{-\frac{N-2}{N}} = J_1(v)\left(\frac{2-N}{2N}J_1(v)\right)^{-\frac{N-2}{N}},$$

that is,

$$J_1(v_\sigma) = \left(\frac{N-2}{2N}\right)^{-\frac{N-2}{N}} (J_1(v))^{2/N}. \tag{15.5}$$

Again, taking into account the P-identity we get

$$\bar{\Phi}(v) = \frac{1}{2}J_1(v) + J_2(v) = \frac{1}{2}J_1(v) + \frac{2-N}{2N}J_1(u) = \frac{1}{N}J_1(v). \tag{15.6}$$

Consequently, by (15.5) and (15.6),

$$\bar{\Phi}(v) = \frac{1}{N}\left(\frac{N-2}{2N}\right)^{\frac{N-2}{2}} (J_1(v_\sigma))^{N/2} > 0. \tag{15.7}$$

Let now v_1 and v_2 from $\mathcal{E}_{p,q}$ be any two nontrivial solutions of (P^∞). For them, $(v_i)_{\sigma_i}$ are taken from formula (15.4), ensuring in each case

$$J_2((v_i)_{\sigma_i}) = -1. \tag{15.8}$$

Taking into consideration the setting of the problem (V^∞) we get (15.3) because for any two nontrivial solutions v_1 and v_2 of problem (P^∞) the inequalities

$$\bar{\Phi}(v_1) \geq \bar{\Phi}(v_2)$$

and

$$J_1((v_1)_{\sigma_1}) \geq J_1((v_2)_{\sigma_2})$$

are equivalent in accordance with (15.7) and (15.8). □

15.3 A General Application

In this section we dwell upon an application of comparison Theorem 15.1. Consider the problem

$$-\Delta u + f(\mathbf{x}, u) = 0 \qquad \text{in } \mathbb{R}^N, \tag{15.9}$$

where f is a Carathéodory function. Denote

$$F(\mathbf{x}, t) = \int_0^t f(\mathbf{x}, s)\, ds.$$

Choose $\mathcal{E}_{p,q}$ as a functional space and consider on it the problem functional Φ for (15.9).

(1) Let $f = f_1 + f_2$, where $f_1 \in K_{p,q}$, $f_2 \in K_q$ with $2 \le p < q < 2^*$, $N > 2$, and $\partial f/\partial u \in K_{p-1\,q-1}$.

(2) Let there be a constant $\theta < 1/2$, positive constants γ_1, γ_2, γ_3, γ_4, and a function $g_0 \in L_1(\mathbb{R}^N)$ such that

$$\begin{aligned} F(\mathbf{x}, t) - \theta\, f(\mathbf{x}, t)\, t &\ge \gamma_1\, |t|^p + g_0(\mathbf{x}), \\ (f_1(\mathbf{x}, t') - f_1(\mathbf{x}, t''))(t' - t'') &\ge \gamma_2\, |t' - t''|^p, \\ F(\mathbf{x}, t) &\ge \gamma_3\, |t|^p - \gamma_4\, |t|^q \end{aligned}$$

for almost all $\mathbf{x} \in \mathbb{R}^N$ and all $t, t', t'' \in \mathbb{R}$.

(3) Let there be a continuous function $\bar{f} : \mathbb{R} \to \mathbb{R}$ such that

$$\lim_{R\to\infty} \sup_{\substack{\mathbf{x}\in\mathbb{R}^N \setminus B_R, \\ t\in\mathbb{R}\setminus\{0\}}} \frac{|f(\mathbf{x}, t) - \bar{f}(t)|}{|t|^{p-1} + |t|^{2^*-1}} = 0.$$

(4) Let there be a $u_1 \in \mathcal{E}_{p,q}$ such that $\Phi(u_1) < 0$.

(5) Let

$$\sup_{0\le t\le 1} \Phi(u_1\, t) < I_0(\bar{\Phi}).$$

Here Φ and $I_0(\bar{\Phi})$ are defined as in Section 15.2.

THEOREM 15.3. *Let all the assumptions of this section hold. Then problem (15.9) has a nontrivial solution $u \in \mathcal{E}_{p,q}$.*

Proof. Comparison Theorem 15.1 implies that if there is no nontrivial solution $u \in \mathcal{E}_{p,q}$ with $\Phi(u) < 0$, then the problem functional satisfies the $\mathrm{PS}_{(0,I_0(\bar{\Phi}))}$-condition.

If a solution u with $\Phi(u) < 0$ exists, then Theorem 15.3 follows immediately. In the opposite case, we shall try to use mountain pass Theorem 13.2. Consider a set $\mathcal{P}$ of paths π connecting points 0 and u_1, where, by (4), $\Phi(u_1) < 0$. By the definition, $\Phi(0) = 0$. As in the proof of Lemma 13.5, we may find a neighborhood of zero of $\mathcal{E}_{p,q}$ such that $\Phi(u) \geq \varepsilon > 0$ on the boundary of this neighborhood. By (5),

$$I \equiv \inf_{\pi \in \mathcal{P}} \sup_{u \in \pi} \Phi(u) \leq \sup_{0 \leq t \leq 1} \Phi(u_1\, t) < I_0(\bar{\Phi}) \tag{15.10}$$

and, consequently,

$$I \in (0, I_0(\bar{\Phi})).$$

All the assumptions of Theorem 13.2 hold, and the existence of a solution $u \in \mathcal{E}_{p,q}$ with $\Phi(u) > 0$ follows. It is nontrivial because $\Phi(0) = 0$. □

Note that condition (5) has an implicit form. We would like to have more transparent conditions. Let us consider some examples.

15.4 The First Example

Theorem 15.4. *Let all the conditions of Theorem 15.3 except for (5) hold. Let there be $\alpha \leq \frac{N-2}{2}$ such that*

$$N\bar{F}(t) - \alpha t \bar{f}(t) \geq 0$$

for all $t \in \mathbb{R}$. Then problem (15.9) has a nontrivial solution $u \in \mathcal{E}_{p,q}$.

Proof. In Section 29.3, we shall prove that any solution $u \in \mathcal{E}_{p,q}$ of (15.9) satisfies functional identity (29.15), which in our case takes the form

$$\left(\frac{N-2}{2} - \alpha\right) \|\nabla u\|_2^2 = -\int_{\mathbb{R}^n} (N\bar{F}(u) - \alpha u \bar{f}(u))\, d\mathbf{x}. \tag{15.11}$$

The left-hand side of (15.11) is nonnegative and the right-hand side by the assumptions of Theorem 15.4 is nonpositive, thus problem (P^∞) associated with (P)=(15.9) has no nontrivial solution in $\mathcal{E}_{p,q}$. Therefore, the definition of I_0 gives $I_0 = +\infty$. Hence, condition (5) of Section 15.3 is valid, and Theorem 15.3 provides the desired result. □

15.5 The Second Example

In this section, we show how to get (15.10) under appropriate conditions in the case where a ground state of the associated problem exists. Consider the problem

$$-\Delta u + a(\mathbf{x})|u|^{p-2}u - b(\mathbf{x})|u|^{q-2}u = 0 \qquad \text{in } \mathbb{R}^N, \tag{15.12}$$

where $N > 2$ and $2 \leq p < q < 2^*$. Let $a, b : \mathbb{R}^N \to \mathbb{R}$ be measurable locally bounded nonnegative functions such that

$$\lim_{R\to\infty} \sup_{\mathbf{x}\in\mathbb{R}^N\setminus B_R} |a(\mathbf{x}) - \bar{a}| = 0,$$

$$\lim_{R\to\infty} \sup_{\mathbf{x}\in\mathbb{R}^N\setminus B_R} |b(\mathbf{x}) - \bar{b}| = 0$$

for some positive constants $\bar{a}$ and $\bar{b}$.

Moreover let

$$\bar{a} \geq a(\mathbf{x}), \qquad \bar{b} \leq b(\mathbf{x})$$

for all $\mathbf{x} \in \mathbb{R}^N$.

THEOREM 15.5. *Let all the above assumptions be satisfied. Then problem (15.12) has a nontrivial solution $u \in \mathcal{E}_{p,q}$.*

Proof. Associate with problem (15.12) the problem

$$-\Delta u + \bar{a}|u|^{p-2}u - \bar{b}|u|^{q-2}u = 0 \qquad \text{in } \mathbb{R}^N. \tag{15.13}$$

It is easy to see that this problem satisfies the conditions of Theorem 9.10 and, by virtue of this theorem and Theorem 15.2, has a ground state $u_0 \in \mathcal{E}_p$, which is a radial function. It is nonnegative by reasoning as in Section 7.8. From the uniqueness theorems of Sections 19.1 and 19.2 it is then strictly positive. If

$$(a(\mathbf{x}) - \bar{a})\, u_0 \equiv (b(\mathbf{x}) - \bar{b})\, u_0 \equiv 0 \tag{15.14}$$

for almost all $\mathbf{x} \in \mathbb{R}^N$, then u_0 satisfies (15.12), and the proof is thus completed.

Assume now (15.14) to be false. Then we continue our reasoning and use Theorem 15.3. Choose $u_0 t_1$ with t_1, large enough, as u_1 in condition (4) of Section 15.3. In accordance with Theorem 15.3 we have only to show that

$$\sup_{0 \leq t < +\infty} \Phi(t\, u_0) < I_0(\bar{\Phi}) = \bar{\Phi}(u_0), \tag{15.15}$$

where

$$\Phi(u) = \frac{1}{2}\|\nabla u\|_2^2 + \frac{1}{p}\|a^{1/p}u\|_p^p - \frac{1}{q}\|b^{1/q}u\|_q^q,$$
$$\bar{\Phi}(u) = \frac{1}{2}\|\nabla u\|_2^2 + \frac{1}{p}\|\bar{a}^{1/p}u\|_p^p - \frac{1}{q}\|\bar{b}^{1/q}u\|_q^q$$

are the problem functionals.

For this purpose consider the behavior of $\bar{\Phi}(t\, u_0)$. We have

$$\frac{\partial \bar{\Phi}(t\, u_0)}{\partial t} = t^{q-1}(t^{2-q}\|\nabla u_0\|_2^2 + \bar{a}\, t^{p-q}\|u_0\|_p^p - \bar{b}\, \|u_0\|_q^q). \tag{15.16}$$

It is easy to see that there is a unique positive value of t at which the derivative (15.16) equals zero. This means that the maximum of $\bar{\Phi}(u_0 t)$ is attained only at a unique positive point t_0. In accordance with equation (15.13)

$$0 = \langle \bar{\Phi}'(u_0), u_0 \rangle = \|\nabla u_0\|_2^2 + \bar{a}\|u_0\|_p^p - \bar{b}\, \|u_0\|_q^q = \left.\frac{\partial \bar{\Phi}(u_0 t)}{\partial t}\right|_{t=1}$$

and, consequently, $t_0 = 1$.

Because $u_0 > 0$ and a, b are nonnegative, we have

$$\sup_{|t|<1/2} \Phi(t\,u_0) \leq \sup_{0\leq t<1/2} \bar{\Phi}(t\,u_0) < \bar{\Phi}(u_0) = I_0(\bar{\Phi})$$

and

$$\sup_{|t|\geq 1/2} \Phi(t\,u_0) < \sup_{|t|\geq 1/2} \bar{\Phi}(t\,u_0) = \bar{\Phi}(u_0) = I_0(\bar{\Phi}).$$

Thus, estimate (15.15) is true under the above assumptions. Hence, the theorem is proved. □

§16. Behavior of PS-sequences in the Symmetric Case. Existence Theorems

We saw earlier in §8 that if an elliptic problem has a radial symmetry, this fact may considerably improve properties of its nontrivial solutions and facilitate its study. In this section not only radial solutions but also slighter variants of symmetries will be investigated. For more information see Lions [4].

16.1 Some Elements of Group Theory

Let $O(\mathbb{N})$ be the group of orthogonal linear transformations of $\mathbb{R}^N$ with natural action. Let G be one of its subgroups. For any point $\mathbf{x} \in \mathbb{R}^N$ the set

$$G_{\mathbf{x}} = \left\{\mathbf{x}_1 \in \mathbb{R}^N : \mathbf{x}_1 = g\mathbf{x},\ g \in G\right\}$$

is called an *orbit* of $\mathbf{x}$. The number of points contained in the orbit $G_{\mathbf{x}}$ will be denoted as $|G_{\mathbf{x}}|$. If this number is infinite, then we write $|G_{\mathbf{x}}| = +\infty$. Denote

$$\#G = \inf_{\mathbf{x}\in\mathbb{R}^N\setminus\{0\}} |G_{\mathbf{x}}|$$

($\#G$ may be, in particular, $+\infty$).

A measurable function f is called *G-symmetric* if for all $g \in G$ and $\mathbf{x} \in \mathbb{R}^N \setminus \{0\}$,

$$f(g\,\mathbf{x}) = f(\mathbf{x})$$

holds.

We consider some examples:

(A) Oddness. Let G_1 consist of two elements $(Id, -Id)$ and act as

$$Id \quad \mathbf{x} = \mathbf{x} \qquad \text{for all } \mathbf{x} \in \mathbb{R}^N,$$
$$(-Id)\,\mathbf{x} = -\mathbf{x} \qquad \text{for all } \mathbf{x} \in \mathbb{R}^N.$$

Any G_1-symmetric function is simply an even function:

$$f(\mathbf{x}) = f(-\mathbf{x}) \qquad \text{for all } \mathbf{x} \in \mathbb{R}^N.$$

We have $\#G_2 = 2$.

(B_1). Let $N > 2$. Consider a group G_2 with elements g_φ, which act as

$$\begin{aligned}
&g_\varphi\,(x^1,\dots,x^N) = (x^1_\varphi,\dots,x^N_\varphi),\\
&x^1_\varphi = -x^2\sin\varphi + x^1\cos\varphi,\\
&x^2_\varphi = x^2\cos\varphi + x^1\sin\varphi,\\
&x^i_\varphi = x^i \qquad i\geq 3,\\
&\varphi\in[0,2\,\pi).
\end{aligned}$$

It is easy to see that

$$\begin{aligned}
&|G_{\mathbf{x}}| = +\infty \qquad &&\text{for } (x^1)^2+(x^2)^2\neq 0,\\
&|G_{\mathbf{x}}| = 1 &&\text{otherwise .}
\end{aligned}$$

Hence $\#G = 1$.

(B_2). Let $N\geq 2$ be even. Let now G_3 act as

$$\begin{aligned}
&g_\varphi\,(x^1,\dots,x^N) = (x^1_\varphi,\dots,x^N_\varphi),\\
&x^i_\varphi = -x^{i+1}\sin\varphi + x^i\cos\varphi, \qquad &&i \text{ is odd},\\
&x^i_\varphi = x^i\cos\varphi + x^{i-1}\sin\varphi, &&i \text{ is even},\\
&\varphi\in[0,2\,\pi).
\end{aligned}$$

It is easy to see that $\#G_3 = +\infty$.

(B_3). Let G_4 be the following subgroup of G_3 acting as in example (B_2): it consists of elements g_φ with $\varphi = \frac{2\pi}{n}k$, where n is an integer and $k = 0,\dots,n-1$. Then $\#G_4 = n$.

(C). Consider the natural action of $O(\mathrm{N})$ on $\mathbb{R}^N\setminus\{0\}$. It is easy to see that in this case, orbits are the spheres ∂B_R, $R>0$, and $\#G = +\infty$. G-symmetric functions are simply the radial functions.

THEOREM 16.1. *Let G act naturally on $\mathbb{R}^N\setminus\{0\}$. Let $1 < \#G < +\infty$. Then there is $\varepsilon > 0$ such that for any $\mathbf{x}\in\partial B_1$ there exist $\#G$ points $\mathbf{x}_i\in G_{\mathbf{x}}$ such that*

$$\inf_{\mathbf{x}_i\neq\mathbf{x}_j}|\mathbf{x}_i-\mathbf{x}_j| > \varepsilon.$$

Proof. Assume the theorem is false. Then for any $\varepsilon>0$ there is an orbit G_ε such that for any finite set of its different points with cardinality $\#G$ there are two different points $\mathbf{x}_\varepsilon$ and $\mathbf{y}_\varepsilon$ such that

$$|\mathbf{x}_\varepsilon-\mathbf{y}_\varepsilon| < \varepsilon.$$

Since the sphere ∂B_1 is compact, one can find an orbit G_0 and a sequence of orbits G_ε such that for any finite $k \geq \#G$ of its points $\mathbf{x}_\varepsilon^i$ there are $\#G-2$ points $\mathbf{z}^i \in G_0$ (some $\mathbf{z}^i$ may coincide with each other) and a point $\mathbf{z}^0 \in G_0$ such that

$$\begin{aligned} \lim_{\varepsilon\to 0} \mathbf{x}_\varepsilon^i &= \mathbf{z}^0, \qquad i = 1,\dots,k-\#G+2,\\ \lim_{\varepsilon\to 0} \mathbf{x}_\varepsilon^i &= \mathbf{z}^i, \qquad i = k-\#G+3,\dots,k. \end{aligned} \tag{16.1}$$

(If $\#G = 2$, $\mathbf{z}^i$ are absent.) Because the natural action of G on $\mathbb{R}^N$ is continuous, it follows from (16.1) that there is an orbit consisting of at most $\#G-1$ points $\mathbf{x}_\varepsilon$ and $\mathbf{z}_\varepsilon^i$. This contradicts the definition of $\#G$. □

16.2 A COMPARISON THEOREM FOR G-SYMMETRIC SEQUENCES

A sequence $\{u_n\}$ of measurable functions $u_n : \mathbb{R}^N \to \mathbb{R}$ will be called *G-symmetric* if all u_n are G-symmetric. For such sequences the results of §15 may be improved.

Recall that we consider the problem

$$-\Delta u + f(\mathbf{x}, u) = 0 \qquad \text{in } \mathbb{R}^N, \tag{P}$$

where f is a Carathéodory function, and the auxiliary problem

$$-\Delta u + \bar{f}(u) = 0 \qquad \text{in } \mathbb{R}^N, \tag{P$^\infty$}$$

with corresponding continuous $\bar{f}$.

THEOREM 16.2. *Let the assumptions of Theorem 14.7 hold. Then if there is no G-symmetric solution $u \in \mathcal{E}_{p,q}$ to problem (P) with $\Phi(u) < 0$, then the functional Φ satisfies the $PS_{(0,\#G\, I_0(\bar{\Phi}))}$-condition on all G-symmetric sequences in $\mathcal{E}_{p,q}$.*

Before proving this theorem we consider the following lemma:

LEMMA 16.3. *Let the assumptions of Theorem 16.2 hold and a sequence $\{u_n\}$ be G-symmetric. Then in the statement of Theorem 14.7, if $\#G < +\infty$, then either $k = 0$ or $k \geq \#G$; if $\#G = +\infty$ then $k = 0$.*

Proof. We consider the proof of Theorem 14.1, which is a simplified version of Theorem 14.7. First, we consider the case $\#G < +\infty$. Let $\{u_n\}$ be a G-symmetric PS_c-sequence of the problem functional for (P) on $\mathcal{E}_{p,q}$. For this sequence the statements of Theorem 14.1 are valid. We suppose that $k > 0$, otherwise Lemma 16.3 is automatically true.

Turn back to the proof of Theorem 14.1, more precisely to relation (I) of §14 which is valid for G-symmetric sequences in $\mathcal{E}_{p,q}$ because it is so for any sequence $\{u_n\} \subset \mathcal{E}_{p,q}$. By virtue of Lemma 14.3, statements (i), (ii), and (iii), the following holds up to a subsequence (cf. (14.37)):

$$\begin{aligned} u_n &\rightharpoonup u_{(0)} \qquad \text{in } \mathcal{E}_{p,q},\\ u_n &\to u_{(0)} \qquad \text{in } \mathcal{E}_{p,q}(B_R) \qquad \text{for any } R, \end{aligned} \tag{16.2}$$

where $u_{(0)}$ is a solution of (P).

It follows that $u_{(0)}$ is a G-symmetric function. Analogously, it follows from definition (14.38),

$$v_n^1 = u_n - u_{(0)},$$

that $\{v_n^1\}$ is a G-symmetric sequence. Recall that we denoted

$$u_{n(1)}(\mathbf{x}) = v_n^1(\mathbf{x} + \mathbf{y}_n^1).$$

Because (14.44) is valid, Lemma 14.4 implies for a subsequence,

$$v_n^1(\,.\,+\mathbf{y}_n^1) \to u_{(1)} \qquad \text{in } \mathcal{E}_{p,q}(B_R) \qquad \text{for any } B_R,$$

where $u_{(1)} \not\equiv 0$ by force of $k \neq 0$.

Because $\{v_n^1\}$ is G-symmetric, it follows for any $g \in G$, $\mathbf{x} \in \mathbb{R}^N$,

$$v_n^1(g(\mathbf{x} + \mathbf{y}_n^1)) = v_n^1(\mathbf{x} + \mathbf{y}_n^1).$$

Then exchanging $g\mathbf{x} \to \mathbf{x}$ we obtain

$$v_n^1(\,.\,+g\mathbf{y}_n^1) \to u_{(1)}(g^{-1}\,.\,) \qquad \text{in } \mathcal{E}_{p,q}(B_R) \text{ for any } B_R.$$

By force of $|\mathbf{y}_n^1| \to \infty$ as $n \to \infty$ and Theorem 16.1, there is a set $\{g_n^i\}$ of $\#G$ elements of the group G such that

$$\min_{i \neq j} \lim_{n\to\infty} |(g_n^i - g_n^j)\mathbf{y}_n^1| \to \infty.$$

Hence, we may find a radius R_0 such that for $\#G$ balls $B_{R_0, g_n^i \mathbf{y}_n^1}$ of the radius R_0 centered at $g_n^i \mathbf{y}_n^1$

$$\varliminf_{n\to\infty} \|v_n^1\|_{\mathcal{E}_{p,q}}(B_{R_0, g_n^i \mathbf{y}_n^1}) \geq \frac{1}{2}\|u_{(1)}\|_{\mathcal{E}_{p,q}}(B_{R_0}) > 0. \tag{16.3}$$

By statement (II), $|g_n^i \mathbf{y}_n^1 - g_n^j \mathbf{y}_n^1| \to \infty$ as $n \to \infty$, and continuity of measure, there is a sequence Ω_n that is a union of k balls of the radii R_n, with $R_n \to \infty$, these balls never intersecting themselves, such that

$$\lim_{n\to\infty} \|v_n^1\|_{\mathcal{E}_{p,q}}(\mathbb{R}^N \setminus \Omega_n) = 0.$$

Now, if $k < \#G$ there are $\#G - k$ balls $B_{R_0, g_n^i \mathbf{y}_n^1} \subset \mathbb{R}^N \setminus \Omega_n$, which contradicts (16.3).

Thus the lemma is proved for the case of finite $\#G$.

If now $\#G$ is infinite, then choosing finite subgroups G_i of G in such a way that $\#G_i < \infty$, we get, as earlier, either $k = 0$ or $k \geq \#G_i$. We may choose finite $\#G_i$ arbitrarily large and $k < +\infty$ by Theorem 14.1, therefore, the only possible case is $k = 0$. □

16.3 PROVING THE COMPARISON THEOREM

Now we are ready to start proving Theorem 16.2 directly.

If there is a solution $u \in \mathcal{E}_{p,q}$ to problem (P) with $\Phi(u) < 0$, then the proof is complete.

Let us now assume the contrary, that is, for any solution $u \in \mathcal{E}_{p,q}$ to (P),

$$\Phi(u) \geq 0 .$$

Consider $c \in (0, \#G\, I_0(\bar{\Phi}))$. By force of Theorem 14.7 we have, analogously to the proof of Theorem 15.1,

$$\#G\, I_0(\bar{\Phi}) > c = \Phi(u_{(0)}) + \sum_{j=1}^{k} \bar{\Phi}(u_{(j)}) \geq \sum_{j=1}^{k} \bar{\Phi}(u_{(j)}) \geq k\, I_0(\bar{\Phi}) . \tag{16.4}$$

Because either $k = 0$ or $k \geq \#G$, (16.4) implies that only $k = 0$ is possible. Then it follows from (I) of §14 that

$$u_n \to u_{(0)} \qquad \text{in } \mathcal{E}_{p,q}.$$

The proof of the theorem is completed. □

16.4 EXAMPLES OF ELLIPTIC PROBLEMS

Here we shall show how the results obtained in the previous sections can be applied to studying the existence of a nontrivial solution to elliptic problems. Instead of considering general problems, we dwell upon a simple example. Consider the problem

$$-\Delta u + u - b(\mathbf{x})|u|^{q-2}u = 0 \qquad \text{in } \mathbb{R}^N, \tag{P}$$

where $N > 2$ and $2 < q < 2^*$. Let $b : \mathbb{R}^N \to \mathbb{R}$ be a measurable locally bounded nonnegative function such that

$$\lim_{R\to\infty} \sup_{\mathbf{x}\in\mathbb{R}^N\setminus B_R} |b(\mathbf{x}) - \bar{b}| = 0 \tag{16.5}$$

for a positive constant $\bar{b}$.

THEOREM 16.4. *Let in addition to condition (16.5) the function b be G_1-symmetric, where G_1 is "the group of evenness" defined in example (A) of Section 16.1, i.e., $b(\mathbf{x}) = b(-\mathbf{x})$, and for almost all $\mathbf{x} \in \mathbb{R}^N$ the inequality*

$$b(\mathbf{x}) > 2^{-\frac{q-2}{2}}\bar{b} \tag{16.6}$$

holds. Then problem (P) has a nontrivial even solution $u \in \mathcal{E}_2$.

THEOREM 16.5. *Let in addition to condition (16.5) the function b be G_3-symmetric, where G_3 is defined in example (B_2) of Section 16.1. Then problem (P) has a nontrivial G_3-symmetric solution $u \in \mathcal{E}_2$.*

16.5 Proving the Existence Theorems

Proof of Theorem 16.4. Denote the subspace of $\mathcal{E}_2$ consisting of functions even in $\mathbf{x}$ as $\mathcal{E}_2^{\text{even}}$. The functional

$$\Phi(u) = \frac{1}{2}\|\nabla u\|_2^2 + \frac{1}{2}\|u\|_2^2 - \frac{1}{q}\|b^{1/q}u\|_q^q$$

is associated with problem (P). If there is a critical point of the functional $\Phi : \mathcal{E}_2^{\text{even}} \to \mathbb{R}$ with $\Phi(u) < 0$, then by force of Appendix C the theorem is proved.

Let now such a point be absent. Then we associate with problem (P) the auxiliary problem

$$-\Delta u + u - \bar{b}|u|^{q-2}u = 0 \qquad \text{in } \mathbb{R}^N. \tag{P$^\infty$}$$

The functional $\bar{\Phi} : \mathcal{E}_2^{\text{even}} \to \mathbb{R}$,

$$\bar{\Phi}(u) = \frac{1}{2}\|\nabla u\|_2^2 + \frac{1}{2}\|u\|_2^2 - \frac{1}{q}\|\bar{b}^{1/q}u\|_q^q,$$

corresponds to problem (P$^\infty$).

According to Theorem 16.2 the functional Φ satisfies the condition $PS_{(0,\#G_1 I_0(\bar{\Phi}))}$ on the space $\mathcal{E}_2^{\text{even}}$. We make use of the mountain pass method. Preliminarily, note that by virtue of Theorem 14.2 problem (P) has a ground state $u_0 \in \mathcal{E}_2$, u_0 being a radial function, that is, in particular, $u_0 \in \mathcal{E}_2^{\text{even}}$.

Keeping in mind Theorem 13.2, choose a family $\mathcal{P}$ of paths $\pi; [0,1] \to X$ with

$$\pi(0) = 0, \qquad \pi(1) = t_0 u_0,$$

where t_0 is a real positive number such that $\Phi(t_0 u_0) < 0$. We saw in Section 13.3 that such a t_0 exists.

It was shown in Section 14.5 that there is a neighborhood U of the origin of $\mathcal{E}_2$, that is, of $\mathcal{E}_2^{\text{even}}$ as well, such that $u_0 t_0 \notin U$ and

$$\inf_{u \in \partial U} \Phi(u) > 0. \tag{16.7}$$

We see that condition (13.2) is valid. We only need to verify that

$$I \in (0, \#G_1 I_0(\bar{\Phi})). \tag{16.8}$$

Assertion (16.18) is a consequence of (16.7) and the following inequality we have to prove

$$I = \inf_{\pi \in \mathcal{P}} \sup_{0 \le t \le 1} \Phi(\pi(t)) < \#G_1 I_0(\bar{\Phi}). \tag{16.9}$$

Because

$$\inf_{\pi \in \mathcal{P}} \sup_{0 \le t \le 1} \Phi(\pi(t)) \le \inf_{\pi \in \mathcal{P}} \sup_{0 \le t \le +\infty} \Phi(t\, u_0), \tag{16.10}$$

it is sufficient to estimate the right-hand side of (16.10) and prove that it is less then $\#G_1 I_0(\bar{\Phi}) = 2\bar{\Phi}(u_0)$. Indeed,

$$\Phi(t\,u_0) = \frac{1}{2}\left(\|\nabla u_0\|_2^2 + \|u_0\|_2^2\right)t^2 - \frac{1}{q}\|b^{1/q}u_0\|_q^q\,t^q. \tag{16.11}$$

Relation (16.11) has a unique maximum in positive t at some t_0 at which $\frac{\partial\Phi(u_0 t)}{\partial t}$ becomes zero. Simple arithmetic gives us this value:

$$t_0 = \left(\frac{\|\nabla u_0\|_2^2 + \|u_0\|_2^2}{\|b^{1/q}u_0\|_q^q}\right)^{1/(q-2)}.$$

Hence,

$$\sup_{0\le t\le+\infty}\Phi(t\,u_0) = \left(\frac{1}{2} - \frac{1}{q}\right)\left(\frac{\|\nabla u_0\|_2^2 + \|u_0\|_2^2}{\|b^{1/q}u_0\|_q^2}\right)^{q/(q-2)}. \tag{16.12}$$

On the other hand, because u_0 is a ground state we have by equation (P^∞),

$$0 = \langle\bar{\Phi}'(u_0), u_0\rangle = \|\nabla u_0\|_2^2 + \|u_0\|_2^2 - \|\bar{b}^{1/q}u_0\|_q^q$$

and, consequently,

$$\bar{\Phi}(u_0) = \left(\frac{1}{2} - \frac{1}{q}\right)\left(\frac{\|\nabla u_0\|_2^2 + \|u_0\|_2^2}{\|\bar{b}^{1/q}u_0\|_q^2}\right)^{q/(q-2)}. \tag{16.13}$$

Comparing (16.12) and (16.13) and taking into account (16.6) we see that the right-hand side of (16.10) is strictly less then $2\bar{\Phi}(u_0)$, hence (16.9) follows. We can now make use of mountain pass Theorem 13.2 and prove the existence of a critical point for the functional Φ. This point is a solution to (P) in accordance with Appendix C. □

Proof of Theorem 16.5. The proof is analogous to that of Theorem 16.4. Choose the subspace of $\mathcal{E}_2$ consisting of G_3-symmetric functions as a functional space $\mathcal{E}_2^{G_3}$ of Theorem 13.2. The problem functional $\Phi : \mathcal{E}_2^{G_3} \to \mathbb{R}$ corresponds to (P). If there is a critical point of $\Phi : \mathcal{E}_2^{G_3} \to \mathbb{R}$ with $\Phi(u) < 0$, then the theorem is proved.

Let now such a point be absent. Then we associate with problem (P) the auxiliary problem (P^∞). The problem functional $\bar{\Phi} : \mathcal{E}_2^{G_3} \to \mathbb{R}$ corresponds to (P^∞). According to Theorem 16.2 the functional Φ satisfies the condition $PS_{(0,\#G_3 I_0(\bar{\Phi}))}$ on the space $\mathcal{E}_2^{G_3}$.

Choose a family $\mathcal{P}$ of paths $\pi : [0,1] \to \mathcal{E}_2^{G_3}$ and a neighborhood U of $\mathcal{E}_2^{G_3}$ analogously to how it was done in proving Theorem 16.4.

It is easy to see analogously to Theorem 16.4 that condition (13.2) is true. We only need to verify

$$I \in (0, \#G_3 I_0(\bar{\Phi})),$$

that is, the inequality

$$\inf_{\pi \in \mathcal{P}} \sup_{u \in \pi} \Phi(u) < \#G_3 \bar{\Phi}(u_0),$$

which is obvious from $\#G_3 = +\infty$.

Thus, the functional $\Phi : \mathcal{E}_2^{G_3} \to \mathbb{R}$ has a nontrivial critical point. This point is a solution to (P). □

§17. Nonradial Solutions of Radial Equations

In the preceding section, G-symmetric solutions for G-symmetric equations were considered. Generally speaking, G-symmetric equations may also have non-G-symmetric solutions. In this section we consider a simple example concerning the radial symmetry.

17.1 SETTING UP THE PROBLEM

Consider the problem

$$-\Delta u + f(u) = 0 \qquad \text{in } \mathbb{R}^N, \tag{17.1}$$

where $N > 2$ and $f : \mathbb{R} \to \mathbb{R}$ is assumed to be a continuous odd function such that (cf. Theorem 9.10)

(a_1) $f(0) = 0$,

(a_2) there exists $t_0 > 0$ such that $F(t_0) < 0$, where

$$F(t) = \int_0^t f(s)\, ds.$$

Let also

(b')

$$0 < \lim_{t \to 0} \frac{F(t)}{|t|^p} < +\infty \qquad \text{for some exponent } 2 \le p < 2^*,$$

(c)

$$0 \le \lim_{t \to \infty} \frac{F(t)}{|t|^{2^*}},$$

(d') $f \in K_{p,2^*}$.

THEOREM 17.1. *Let $N = 4$ or be greater than 5. Let an odd continuous f satisfy conditions (a), (b'), (c), and (d'). Then problem (17.1) has a nonradial solution $u \in \mathcal{E}_p$.*

In order to prove this theorem we consider new functional spaces.

17.2 Block-Radial Symmetry

Any point $\mathbf{x} \in \mathbb{R}^N$ is fixed by its coordinates

$$(x_1, \dots, x_N).$$

For a given set of m natural numbers $\mathcal{N}_i$ satisfying the relations

$$\mathcal{N}_i \geq 2 \qquad \text{for all } i = 1, \dots, m,$$
$$\sum_{i=1}^{m} \mathcal{N}_i = N,$$

define functions $s_i : \mathbb{R}^{\mathcal{N}_i} \to \mathbb{R}^+$ as

$$s_1 = s(x_1, \dots, x_{\mathcal{N}_1}) = (x_1^2 + \dots + x_{\mathcal{N}_1}^2)^{1/2},$$
$$s_2 = s(x_{\mathcal{N}_1+1}, \dots, x_{\mathcal{N}_1+\mathcal{N}_2}) = (x_{\mathcal{N}_1+1}^2 + \dots + x_{\mathcal{N}_1+\mathcal{N}_2}^2)^{1/2},$$
$$\dots\dots\dots\dots\dots\dots\dots\dots\dots\dots\dots\dots\dots\dots,$$
$$s_m = s(x_{\mathcal{N}_{m-1}+1}, \dots, x_N) = (x_{\mathcal{N}_{m-1}+1}^2 + \dots + x_N^2)^{1/2}.$$

DEFINITION. A function $u \in \mathcal{E}_{p,q}$ is called *block-radial* if it depends only on s_i.

Corresponding spaces of such functions with a prescribed set $(s_1, \dots, s_m)$ will be denoted as $\mathcal{E}_{p,q}^{(s_1,\dots,s_m)}$, $\mathcal{E}_p^{(s_1,\dots,s_m)}$, or $L_q^{(s_1,\dots,s_m)}$, respectively.

We shall investigate the solvability of problem (17.1) in the spaces $\mathcal{E}_{p,q}^{(s_1,\dots,s_m)}$. The following important result is true.

THEOREM 17.2. *Let $2 \leq p < q < 2^*$. Then*

$$\mathcal{E}_p \circlearrowleft\!\circlearrowright L_q^{(s_1,\dots,s_m)}.$$

Proof. If $m = 1$, $\mathcal{E}_p^{(s_1,\dots,s_m)}$ is the space of radial functions in $\mathbb{R}^N$ with $N \geq 2$, and Theorem 17.2 for this case was proved in Section 8.2 (see Lemma 8.5).

Other cases are more cumbrous for proving. We refer to the papers of Lions [1], Ding Yi [1], and Kuzin [1], where Theorem 17.2 is proved with the help of three different approaches. □

17.3 Completing the Proof

Here we justify Theorem 17.1.

Let N be as in the assumptions of Theorem 17.1. Choose two different indexes i and j and assume $\mathcal{N}_i = \mathcal{N}_j$. Denote a subspace of $\mathcal{E}_{p,q}^{(s_1,\dots,s_m)}$ consisting of the functions u with

$$u(s_1, \dots, s_i, \dots, s_j, \dots, s_m) = -u(s_1, \dots, s_j, \dots, s_i, \dots, s_m)$$

as $\mathcal{E}_{p,q}^{(s_1,\dots,[s_i,s_j],\dots,s_m)}$.

THEOREM 17.3. *Let the assumptions of Theorem 17.1 be true. Then for any space* $\mathcal{E}_{p,q}^{(s_1,\dots,[s_i,s_j],\dots,s_m)}$ *there is a nontrivial solution* $u \in \mathcal{E}_{p,q}^{(s_1,\dots,[s_i,s_j],\dots,s_m)}$ *for problem (17.1).*

Proof. We associate with problem (17.1) the eigenvalue problem

$$\begin{aligned} -\Delta u + \lambda f(u) = 0 \qquad & \text{in } \mathbb{R}^N, \\ u \to 0 \qquad \text{as } |\mathbf{x}| \to \infty. & \end{aligned} \tag{17.2}$$

In full analogy to §12 we conclude that if $u \in \mathcal{E}_{p,q}^{(s_1,\dots,[s_i,s_j],\dots,s_m)}$ is a solution to (17.2) with $\lambda > 0$, then $u(\,.\,/\sqrt{\lambda})$ is a solution to (17.1). In order to establish the existence result for (17.2) we make use of the variational method (cf. §9). Consider the problem

$$\left\{u \in \mathcal{E}_{p,q}^{(s_1,\dots,[s_i,s_j],\dots,s_m)},\ \min J_1(u),\ J_2(u) = -1\right\}, \tag{17.3}$$

where

$$J_1(u) = \|\nabla u\|_2^2,$$
$$J_2(u) = \int\limits_{\mathbb{R}^N} F(u)\, d\mathbf{x},$$

and

$$F(t) = \int\limits_0^t f(s)\, ds.$$

If u is a solution to problem (17.3), then by the Lagrange theorem and reasons analogous to those of Theorem 6.6 there is a $\lambda > 0$ such that

$$\langle J_1'(u), \varphi\rangle + \lambda\, \langle J_2'(u), \varphi\rangle = 0 \tag{17.4}$$

for all $\varphi \in \mathcal{E}_{p,q}^{(s_1,\dots,[s_i,s_j],\dots,s_m)}$. From the results of Appendix C, Theorem C.4, it follows that (17.4) remains true for all $\varphi \in \mathcal{E}_{p,q}$ as well, and thus u is a solution to (17.2).

Now we only need to get the existence result for the variational problem (17.2). The proof almost repeats that of Theorem 9.10. The only complication concerns the following step. □

LEMMA 17.4. *The set*

$$\left\{u \in \mathcal{E}_{p,q}^{(s_1,\dots,[s_i,s_j],\dots,s_m)},\ J_2(u) = -1\right\}$$

is not empty.

Proof. Consider a family of functions $v_R : (\mathbb{R}^+)^m \to \mathbb{R}$ of the form $v_R = w_R\varphi$, where $w_R : (\mathbb{R}^+)^m \to \mathbb{R}$ has the form

$$w_R(r) = \begin{cases} t_0, & r \le R, \\ t_0(R - r + 1), & R < r \le R+1, \\ 0, & r \ge R+1, \end{cases}$$

$r = (s_1^2 + \cdots + s_m^2)^{1/2}$, and φ has the form

$$\varphi(\mathbf{s}) = \begin{cases} -1, & s_j + 2 \le s_i, \\ s_j - s_i + 1, & s_j + 1 < s_i \le s_j + 2, \\ 0, & s_j - 1 < s_i \le s_j + 1, \\ s_j - s_i - 1, & s_j - 2 < s_i \le s_j - 1, \\ 1, & s_i < s_j - 2. \end{cases}$$

It is easy to see that $v_R \in \mathcal{E}_{p,q}^{(s_1,\ldots,[s_i,s_j],\ldots,s_m)}$. For these functions we have

$$\begin{aligned} J_2(v_R) = S_N \int_0^R F(t_0) r^{N-1}\, dr &+ \int_{B_{R+1}\setminus B_R} F(t_0(R + r - 1)\varphi(\mathbf{s}))\, d\mathbf{x} \\ &+ \int_{B_R \cap \{s_j - 2 \le s_i \le s_j + 2\}} (F(t_0\varphi(\mathbf{s})) - F(t_0))\, d\mathbf{x}. \end{aligned}$$

Estimate these integrals:

$$\int_0^R F(t_0) r^{N-1}\, dr = \frac{1}{N} F(t_0)\, R^N,$$

$$\left| \int_{B_{R+1}\setminus B_R} F(t_0(R + r - 1)\varphi(\mathbf{s}))\, d\mathbf{x} \right| \le S_N \max_{0 \le t \le t_0} |F(t)| \left((R+1)^N - R^N\right)/N,$$

$$\int\limits_{B_R\cap\{s_j-2\le s_i\le s_j+2\}} (F(t_0\varphi(\mathbf{s}))-F(t_0))\,d\mathbf{x}$$

$$\le \mathbf{C}\max_{0\le t\le t_0}|F(t)| \int\limits_{(\mathbb{R}^+)^3\cap\{s_i^2+s_j^2+t^2\le R\}\cap\{s_j-2\le s_i\le s_j+2\}} (s_is_j)^{\mathcal{N}_i-1}t^{N-2\mathcal{N}_i-1}\,ds_i\,ds_j\,dt$$

$$\le \mathbf{C}\,R^{N-2\mathcal{N}_i} \int\limits_{(\mathbb{R}^+)^2\cap\{t_1^2+t_2^2\le R^2\}\cap\{|t_2|\le 4\}} |t_1^2-t_2^2|^{\mathcal{N}_i-1}dt_1\,dt_2$$

$$\le \mathbf{C}\,R^{N-2\mathcal{N}_i}\int_0^R |t_1^2+64|^{\mathcal{N}_i-1}dt_1 \le \mathbf{C}+\mathbf{C}\,R^{N-1},$$

where the different positive constants $\mathbf{C}$ do not depend on R.

It can be easily seen that

$$\frac{J_2(v_R)}{R^N}\to\frac{1}{N}F(t_0)\,S_N<0$$

as $R\to+\infty$, and, consequently, there is a radius R_0 and a positive constant ε such that

$$J_2(v_{R_0})=-\varepsilon_0.$$

By scaling Lemma 9.2,

$$J_2(v_{R_0}(\,.\,\varepsilon_0^{1/N}))=-1.$$

Thus, Lemma 17.4 and hence Theorem 17.1 are proved. □

§18. Methods of Bounded Domains Approximation

As mentioned previously, one of the main difficulties arising in the study of elliptic problems in $\mathbb{R}^N$ originates from the fact that the embedding $\mathcal{E}_{2^*}\circlearrowright L_p(\mathbb{R}^N)$ with $1<p<2^*$ is not compact. For bounded domains this embedding is compact, which facilitates the investigation of variational problems. According to the results of §3 we, consequently, may try to approximate a solution of a noncoercive problem in $\mathbb{R}^N$ by solutions of appropriate problems in B_R as $R\to\infty$. The investigation of the problems in B_R is made by variational methods described earlier. This section is devoted to an example of this kind due to Ding and Ni [1].

18.1 SETTING UP THE PROBLEM

Consider the problem

$$-\Delta u+u-b(|\mathbf{x}|)|u|^{q-2}u=0 \qquad \text{in } \mathbb{R}^N, \tag{P$^\infty$}$$

where $N>2$, $2<q<2^*$, and $b:\mathbb{R}^N\to\mathbb{R}$ is a nonnegative nontrivial locally Hölder function. Denote $r=|\mathbf{x}|$.

THEOREM 18.1. *Let there be a constant $C > 0$ and an exponent l such that*

$$0 < l < \frac{1}{2}(N-1)(q-2) \tag{18.1}$$

and for all $\mathbf{x} \in \mathbb{R}^N$, the inequality

$$0 \le b(\mathbf{x}) \le C\,(1 + r^l) \tag{18.2}$$

holds. Then problem (P^∞) has a nontrivial solution $u \in H_0^1(\mathbb{R}^N)$.

In order to prove Theorem 18.1 we associate with problem (P^∞) the eigenvalue problem

$$-\Delta u + u - \lambda\, b(|\mathbf{x}|)|u|^{q-2}u = 0 \qquad \text{in } \mathbb{R}^N. \tag{P_λ^∞}$$

By force of the homogeneity of the equation (see §12), Theorem 18.1 is a consequence of

THEOREM 18.2. *Let the conditions of Theorem 18.1 hold. Then there exist a constant $\lambda > 0$ and a nontrivial $u \in H_0^1(\mathbb{R}^N)$ that solve (P_λ^∞).*

Proof. We make use of the methods of §3, Theorem 3.3. For this purpose let us study the problem in the balls $B_R \subset \mathbb{R}^N$.

18.2 EXISTENCE RESULTS FOR BALLS

Consider a family of problems

$$\begin{aligned} &-\Delta u + u - \lambda\, b(|\mathbf{x}|)|u|^{q-2}u = 0 \qquad \text{in } B_R, \\ &u(\mathbf{x}) = 0 \qquad \text{for } \mathbf{x} \in \partial B_R. \end{aligned} \tag{P_λ^R}$$

THEOREM 18.3. *Let the conditions of Theorem 18.2 hold. Let $b \not\equiv 0$ in B_{R_0}. Then for any $R \ge R_0$, problem (P_λ^R) has a solution $(\lambda, u_R) \in \mathbb{R} \times H_0^1(\mathbb{R}^N)$ with $\lambda > 0$ and nonnegative $u \not\equiv 0$.*

Proof. We consider the variational problem

$$\{u \in H_0^{1,\mathrm{rad}}(B_R),\ \min J_1(u),\ J_2(u) = 1\},$$

where functionals $J_1, J_2 : H_0^{1,\mathrm{rad}}(B_R) \to \mathbb{R}$ have the form

$$J_1(u) = \|\nabla u\|_2^2(B_R) + \|u\|_2^2(B_R),$$
$$J_2(u) = \|b^{1/q}u\|_q^q(B_R),$$

and $H_0^{1,\mathrm{rad}}(B_R)$ are the subspaces of $H_0^1(B_R)$ consisting of radial functions.

As in §7, we may use Theorem 6.3 and prove that problem (P_λ^R) has a solution with the required properties. Note that the only difference in the proof consists in the establishing of weakly lower continuousness of the functional J_2. Because b is uniformly bounded on every finite ball, the compact embedding $H_0^1(B_R) \circlearrowright\circlearrowright L_q(B_R)$ guaranteed by the Kondrashov theorem ensures this fact. □

18.3 BOUNDEDNESS OF SEQUENCES

Here we verify condition (3.7).

LEMMA 18.4. *There are positive constants* $\mathbf{C}$ *not depending on* R *and* u *such that for all* $R > R_0$,

$$\|\nabla u_R\|_2 + \|u_R\|_2 < \mathbf{C}, \tag{18.3}$$

$$|\lambda_R| < \mathbf{C}. \tag{18.4}$$

Proof. As mentioned in §3 any solution $u_R \in H_0^1(B_{R_1})$ may be considered as a function in $H_0^1(\mathbb{R}^N)$ or $H_0^1(B_{R_2})$, provided $R_2 > R_1$. Hence

$$\inf_{\substack{u\in H_0^1(B_{R_2})\\ J_2(u)=1}} J_1(u) \le \inf_{\substack{u\in H_0^1(B_{R_1})\\ J_2(u)=1}} J_1(u)$$

for all $R_2 \ge R_1$. Then for $R \ge R_0$,

$$\|\nabla u_R\|_2^2 + \|u_R\|_2^2 = J_1(u_R) \le J_1(u_{R_0}) \le \mathbf{C}.$$

Property (18.3) is proved. Now we turn to (18.4). Choose u_R as a test function v in the identity

$$\langle J_1'(u), v\rangle - \lambda_R \langle J_2'(u_R), v\rangle = 0.$$

Then taking into account the explicit forms of the functionals J_1, J_2 and their derivatives, we get

$$\begin{aligned}\lambda_R = \lambda_R\, J_2(u_R) &= \frac{\lambda_R}{q}\langle J_2'(u_R), u_R\rangle \\ &= \frac{1}{q}\langle J_1'(u_R), u_R\rangle = \frac{2}{q}J_1(u_R) \le \frac{2}{q}J_1(u_{R_0}) = \mathbf{C}\,.\end{aligned}$$

Moreover, as we have just seen,

$$\lambda_R = \frac{2}{q}J_1(u_R) \ge 0 \tag{18.5}$$

and (18.4) follows. □

18.4 PROOF OF THEOREM 18.2

Now we can complete the proof of Theorem 18.2. According to Theorem 18.3 and the properties of solutions stated in Lemma 18.4 there exists a sequence $\{R_i\}$ such that

$$u_{R_i} \rightharpoonup u \qquad \text{in } H_0^1(\mathbb{R}^N),$$
$$\lambda_{R_i} \to \lambda \qquad \text{in } \mathbb{R},$$

where $u_{R_i} \not\equiv 0$ and $\lambda_{R_i} > 0$ form a solution to $(\mathrm{P}_\lambda^{R_i})$. To finish the proof we need only to show that both $u \not\equiv 0$ and $\lambda \neq 0$.

Note that by force of radial Lemma 8.1 and (18.3) the estimate

$$|u_R(r)| \leq \mathbf{C}\, \|\nabla u_R\|_2\, r^{\frac{1-N}{2}} \leq \mathbf{C}\, r^{\frac{1-N}{2}} \tag{18.6}$$

holds. (Here and later on constants $\mathbf{C}$ are assumed not to depend on R, and for the sake of simplicity indexes i for sequences $\{R_i\}$ are omitted.)

By regularity Theorem 1.4, we have $u_R \in C^2(\mathbb{R}^N)$. Moreover, by (18.6), $u_R(r)$ tends uniformly to zero as $r \to \infty$. By Section 7.8 we may choose $u_R \geq 0$. Hence there exists a point r_R such that $u_R(r_R)$ is a positive maximum of u_R on $[0, +\infty)$. Then $u_R'(r_R) = 0$ and $u_R''(r_R) \leq 0$. Consequently, we have by the equation

$$u_R(r_R) - \lambda_R b(r_R)(u_R(r_R))^{q-1} \leq 0$$

or, equivalently,

$$u_R \geq (\lambda_R b(r_R))^{-1/(q-2)}. \tag{18.7}$$

From (18.6), (18.2), and (18.7) the estimate follows:

$$\mathbf{C} \geq u_R(r_R)\, r_R^{\frac{N-1}{2}} \geq \mathbf{C}\, \lambda_R^{-1/(q-2)} r_R^{\frac{N-1}{2}} (1 + r_R^l)^{-1/(q-2)}. \tag{18.8}$$

Because λ_R are uniformly bounded, we have, taking into account (18.1) and (18.8),

$$r_R \leq R_1 \tag{18.9}$$

for some $R_1 \geq R_0$. Due to (18.8) and (18.9), we have

$$u_R(r_R) \geq \mathbf{C}\, \lambda_R^{-1/(q-2)} (1 + r_R^l)^{-1/(q-2)} \geq \mathbf{C}\, \lambda_R^{-1/(q-2)} (1 + R_1^l)^{-1/(q-2)}, \tag{18.10}$$

that is, a constant C_0 exists such that for all $R \geq R_1$,

$$\max_{\mathbf{x} \in \mathbb{R}^N} u_R(\mathbf{x}) \geq C_0. \tag{18.11}$$

By regularity theorems analogous to those of Section 1.2 (see the remark in Section 3.1) and estimate (18.3),

$$\max_{\mathbf{x} \in B_{R_1}} |\nabla u_R(\mathbf{x})| \leq \mathbf{C}. \tag{18.12}$$

Relations (18.11) and (18.12) lead to the existence of a constant $\varepsilon > 0$ such that, for all $R > R_1$,

$$\|\nabla u_R\|_2^2(B_{R_1}) + \|u_R\|_2^2(B_{R_1}) \geq \varepsilon > 0. \tag{18.13}$$

According to the Kondrashov theorem we have

$$u_{R_i} \to u \qquad \text{in } H_0^1(B_{R_1}).$$

Hence, (18.13) implies $u \not\equiv 0$.

From (18.5) and (18.13), it follows that

$$\lambda_R = \frac{2}{q} J_1(u_R) \geq 2\varepsilon/q > 0$$

for all $R \geq R_1$. Hence, $\lambda \neq 0$. □

Chapter 4
Radial Solutions: The ODE Method

In this chapter, we consider the equation

$$-\Delta u + f(r, u) = 0 \qquad \text{in } \mathbb{R}^N, \tag{19.1}$$

where we denote $r = |\mathbf{x}|$.

§19. Basic Techniques of the ODE Method

Because the function f depends only on r, we may search for solutions in the class of radial functions. In this case equation (19.1) takes the form

$$u'' + \frac{N-1}{r}u' - f(r, u) = 0, \qquad r \geq 0, \tag{19.2}$$

where the prime denotes differentiation with respect to r. Equation (19.2) may be investigated by the methods of ordinary differential equations (ODE methods). We shall be interested only in classical solutions. The requirement $u \in C^2_{\text{loc}}(\mathbb{R}^N)$ in (19.1) implies the initial condition $u'(0) = 0$ in (19.2). We may also impose an initial condition on the value of the function u itself and study the behavior of solutions with regard to $u(0)$. Thus, an investigation of problem (19.1) in the radial case can be reduced to the study of the Cauchy problem

$$\begin{aligned} &u'' + \frac{N-1}{r}u' - f(r, u) = 0, \qquad r \geq 0, \\ &u(0) = a, \\ &u'(0) = 0. \end{aligned} \tag{19.3}$$

19.1 Local Properties

Here and later on in this chapter, we always suppose that the function $f : (r, t) \mapsto f$ is continuous in r for all t and locally Lipschitz in t for all r. The following is well known.

THEOREM 19.1. *Under the above conditions, given a, there exists a constant $\varepsilon > 0$ such that the problem (19.3) has a unique solution $u \in C^2([0, \varepsilon))$.*

We also may formulate this theorem for initial data given at a point different from the origin.

THEOREM 19.2. *Under the above conditions, given $R > 0$ and real a, b, there exists a constant $\varepsilon > 0$ such that the problem*

$$\begin{aligned} &u'' + \frac{N-1}{r}u' - f(r,u) = 0, \qquad r \geq R, \\ &u(R) = a, \\ &u'(R) = b, \end{aligned} \tag{19.4}$$

has a unique solution $u \in C^2([R, R+\varepsilon))$.

We denote the solution of (19.3) with fixed a by u_a, analogously we denote the solution to (19.4) as $u_{a,b}$. Solutions of problem (19.3) or (19.4) depend continuously on the initial data:

THEOREM 19.3. *For any $R_0 > 0$, $\delta > 0$, and real a there exists $\varepsilon > 0$ such that if u_a and $u_{\bar{a}}$ are $C^2([0, R_0])$ solutions of (19.3) with*

$$|a - \bar{a}| < \varepsilon,$$

then

$$\max_{0 \leq r \leq R_0} (|u_a(r) - u_{\bar{a}}(r)| + |u'_a(r) - u'_{\bar{a}}(r)|) < \delta.$$

THEOREM 19.4. *For any $R_0 > R > 0$, $\delta > 0$, and real a, b there exists $\varepsilon > 0$ such that if $u_{a,b}$ and $u_{\bar{a},\bar{b}}$ are $C^2([R, R_0])$ solutions of (19.4) with*

$$|a - \bar{a}| + |b - \bar{b}| < \varepsilon,$$

then

$$\max_{R \leq r \leq R_0} (|u_{a,b}(r) - u_{\bar{a},\bar{b}}(r)| + |u'_{a,b}(r) - u'_{\bar{a},\bar{b}}(r)|) < \delta.$$

The following is also used often. Denote by $u_{(i)}$, $i = 1, 2$, the solutions of

$$\begin{aligned} &u'' + \frac{N-1}{r}u' - f(r,u) = h_{(i)}(r), \qquad r \geq 0, \\ &u(0) = a, \\ &u'(0) = 0, \end{aligned}$$

where $h_{(i)} : \mathbb{R}^+ \to \mathbb{R}$ are some continuous functions.

THEOREM 19.5. *For any $R > 0$, $\delta > 0$, and real a there exists $\varepsilon > 0$ such that if*

$$\max_{0 \leq r \leq R} |h_{(1)}(r) - h_{(2)}(r)| < \varepsilon,$$

then the $C^2([0, R])$ solutions u_1 and u_2, if exist, satisfy

$$\max_{0 \leq r \leq R} (|u_{(1)}(r) - u_{(2)}(r)| + |u'_{(1)}(r) - u'_{(2)}(r)|) < \delta.$$

Proof of theorems. We only outline the scheme for equation (19.3) with $N > 2$. Equation (19.3) may be rewritten as

$$(u'r^{N-1})' = f(r,u)r^{N-1}. \tag{19.5}$$

Integration gives

$$u'(r) = r^{-(N-1)} \int_0^r f(t,u(t))t^{N-1}\,dt. \tag{19.6}$$

The second integration together with a change of the order of integration gives

$$u(r) = \frac{1}{N-2} \int_0^r t\left(1 - \left(\frac{t}{r}\right)^{N-2}\right) f(t,u(t))\,dt + a. \tag{19.7}$$

The fact that (19.5) follows from (19.7) may be verified directly by differentiation.

With the help of the ordinary technique of contracting mappings (e.g., Fucik and Kufner [1]) the local existence of $u \in C([0,\varepsilon))$ and local uniqueness of a solution to (19.7) may be proved.

That $u \in C^2([0,\varepsilon))$ follows from (19.7), conditions on f, and $u \in C([0,\varepsilon))$. □

Remark. All procedures are standard except maybe those that are caused by the singularity at the origin. But this difficulty can be easily overcome. For example, equation (19.6) yields, by change of variables,

$$\frac{u'(r)}{r} = \int_0^1 f(rt,u(rt))t^{N-1}\,dt.$$

Then, taking into consideration the continuousness of f, we have

$$\lim_{r\to+0} \frac{u'(r)}{r} = \frac{f(0,u(0))}{N}.$$

The proof of the continuous dependence of solutions to (19.4) on the initial data are presented practically in all books on ordinary differential equations. If $R = 0$ and $b = 0$, we come to the problem (19.3), which can be treated analogously with the slightest modifications.

For more detailed information on ordinary differential equations see Pontryagin [1] and Sansone [1].

19.2 Extension of Solutions. Entire Solutions

When we investigated problem (19.1) by variational methods, they produced solutions naturally defined on $\mathbb{R}^N$. In the approach based on the Cauchy problem the situation is different. A local solution to (19.3) may blow up, i.e., tend to infinity at some finite r. We shall be interested in the existence of global solutions defined for all $r \in [0, +\infty)$. Such solutions are called entire. In order to investigate the relationship between local and global solutions we consider possible extensions of local ones.

DEFINITION. Let a function $u \in C^2([0, \varepsilon))$ satisfy (19.3). Then a function $v \in C^2([0, \varepsilon_0))$, with $\varepsilon_0 > \varepsilon$, satisfying (19.3) such that

$$v(r) = u(r) \qquad \text{for } r \in [0, \varepsilon)$$

is called an extension of u.

Since we assumed that f satisfies the regularity conditions, an extension, if it exists, is unique by Theorem 19.1 or 19.2. The following theorem is known (e.g., Sansone [1]).

THEOREM 19.6. *Let $f(r, u)$ be a function locally Hölder in the first and locally Lipschitz in the second variable. Let a function $u_a \in C^2([0, \varepsilon])$ be a solution of (19.3). Then it has a unique extension that either*

(1) *is entire, or*
(2) *is such that there is a point r_0 where*

$$\overline{\lim_{r \to r_0 - 0}} |u_a(r)| = \infty,$$
$$\overline{\lim_{r \to r_0 - 0}} |u_a'(r)| = \infty.$$

With the help of these preliminary facts, we may start studying the problem of global solvability itself.

§20. Autonomous Equations in the N-dimensional Case

In this paragraph we consider the global solvability of the problem

$$\begin{aligned} &u'' + \frac{N-1}{r} u' - f(u) = 0, \qquad r \geq 0, \\ &u(0) = a, \\ &u'(0) = 0, \end{aligned} \tag{20.1}$$

where f does not depend on r. The function f is supposed to be locally Lipschitz-continuous. Because we always may redefine $f(\,.\, - a) = f_1(\,.\,)$ we suppose without loss of generality that $a \geq 0$. From Theorem 19.1 problem (20.1) has a local solution, which will be denoted as u_a. We shall investigate whether it admits an entire extension.

20.1 "ANTICOERCIVE" EQUATIONS

Denote

$$F(t) = \int_0^t f(s)\, ds.$$

THEOREM 20.1. *Let there exist a constant $C > 0$ such that the inequality*

$$F(t) \leq C \tag{20.2}$$

holds for all $t \in \mathbb{R}$. Then for any $a \geq 0$ the extended solution u_a is entire.

If, additionally,

$$\lim_{|t|\to\infty} F(t) = -\infty, \tag{20.3}$$

the extended solution u_a is uniformly bounded on $\mathbb{R}^+$.

Proof. We follow the method of Walter [1]. Let u_a be a local solution to problem (20.1). Multiply equation (20.1) by u_a' and integrate from 0 to R. We get, taking into consideration $u_a'(0) = 0$ and $F(u_a(0)) = F(a)$,

$$\frac{1}{2}|u_a'(R)|^2 + (N-1)\int_0^R \frac{|u_a'(r)|^2}{r}\, dr + F(a) - F(u_a(R)) = 0. \tag{20.4}$$

Define the function $E : \mathbb{R}^+ \to \mathbb{R}$ by

$$E(R) = \frac{1}{2}|u_a'(R)|^2 - F(u_a(R)). \tag{20.5}$$

From (20.1) we have

$$E'(R) = -\frac{N-1}{R}|u_a'(R)|^2 \leq 0.$$

Therefore, E is nonincreasing and

$$E(R) \leq E(0) = -F(a). \tag{20.6}$$

Consequently, from (20.2), (20.5), and (20.6),

$$\frac{1}{2}|u_a'(R)|^2 \leq -F(a) + F(u_a(R)) \leq C - F(a).$$

Because u_a' is bounded, a local solution may be extended to an entire solution by Theorem 19.6.

Now (20.5) and (20.6) imply $F(u_a(R)) \geq F(a)$. If (20.3) holds, then the boundedness of $F(u_a)$ from below yields the boundedness of u_a. □

20.2 EQUATIONS WITH A POTENTIAL BARRIER

THEOREM 20.2. *Let there exist a constant C such that the equation*

$$F(s) = C$$

in s has two solutions $\alpha < \beta$ with

$$F(a) > C \tag{20.7}$$

for all $a \in (\alpha, \beta)$. Then for all $a \in (\alpha, \beta)$ the solution u_a to (20.1) is entire and bounded by α and β.

Proof. By the assumption of the theorem,

$$F(\alpha) = F(\beta) = C. \tag{20.8}$$

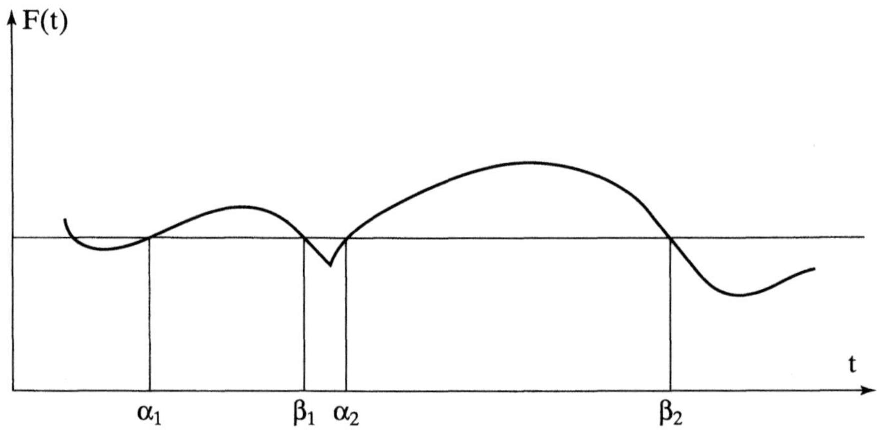

Fig. 20.1

See as an example Fig. 20.1. Fix some $a \in (\alpha, \beta)$. If $f(a) = 0$, equation (20.1) has the entire bounded solution $u_a \equiv a$. If $f(a) \neq 0$, then $u_a \not\equiv$ Const and, from (20.4), we get

$$F(u_a(R)) = \frac{1}{2}|u_a'(R)|^2 + (N-1)\int_0^R \frac{|u_a'(r)|^2}{r}\,dr + F(a) \geq F(a). \tag{20.9}$$

We claim that

$$\alpha < u_a(r) < \beta.$$

Indeed, because $\alpha < u_a(0) < \beta$, $u_a(r)$ may be equal to α or β only at a point $R > 0$ where, by (20.8) and (20.9),

$$C = F(u_a(R)) \geq F(a),$$

which contradicts (20.7). □

20.3 "WEAKLY COERCIVE" EQUATIONS

THEOREM 20.3. *Let for all $t > 0$ the inequality $f(t) > 0$ hold and*

$$\overline{\lim_{t\to+\infty}} \frac{F(t)}{t^2} \le m^2 \tag{20.10}$$

with $m > 0$. Then for any $a > 0$ the solution u_a is positive and entire.

Proof. From (19.6) and the condition $f > 0$, we have $u_a'(r) > 0$ for $r > 0$. Note that (20.4) and (20.10) provide the estimate

$$\frac{1}{2}|u_a'(R)|^2 \le -F(a) + F(u_a(R)) \le 2m^2|u_a(R)|^2 + C_a \tag{20.11}$$

with some constant $C_a > 0$. Because $u_a' > 0$ and $u_a > 0$, the inequality (20.11) may be rewritten as

$$u_a' - 2mu_a \le \text{Const}\,.$$

Hence,

$$e^{2mr}(e^{-2mr}u_a(r))' \le \text{Const}\,.$$

Integrating, we see that there exists a constant C_0 such that

$$u_a(r) \le C_0\, e^{2mr}. \tag{20.12}$$

Estimate (20.12) and Theorem 19.6 complete the proof. □

20.4 "STRONGLY COERCIVE" EQUATIONS

We shall show that the conditions in previous sections are in some sense unimprovable.

THEOREM 20.4. *Let $f(t) > 0$ for all $t > 0$ and $p > 1$ such that*

$$\lim_{t\to+\infty} \frac{f(t)}{t^p} = m^2 > 0. \tag{20.13}$$

Then for all $a > 0$, the solution u_a can not be extended onto $[0, +\infty)$.

Proof. First step. Let the theorem be not true and u_a be an entire solution for some $a > 0$. Then from the equation we conclude analogously to Section 20.3 that

$$u_a(r) > 0, \quad u_a'(r) > 0 \qquad \text{for } r > 0. \tag{20.14}$$

We claim that for any C there is an R such that

$$u_a(R) > C. \tag{20.15}$$

Indeed, if (20.15) were not true, (20.14) would imply that there is a constant $C_a > 0$ such that

$$\begin{aligned}\lim_{r\to\infty} u_a(r) &= C_a,\\ \lim_{r\to\infty} u_a'(r) &= 0.\end{aligned} \tag{20.16}$$

The equation then yields

$$\lim_{r\to\infty} u_a''(r) = f(C_a) > 0,$$

which contradicts (20.16).

Analogously to (20.15), for any C there is an R such that

$$u_a'(R) > C. \tag{20.17}$$

Assertion (20.15) and assumption (20.13) provide the existence of R such that

$$\Delta u_a \geq \frac{1}{2} u_a^p m^2 \tag{20.18}$$

holds for all $r \geq R$. □

Second step. Denote

$$\begin{aligned}u_a(R) &= C_1,\\ u_a'(R) &= C_2,\\ \alpha &= 2/(p-1),\\ \lambda &= (4\alpha(\alpha+1)/m^2)^{\alpha/2}.\end{aligned}$$

Taking into consideration (20.14) and (20.17) we may choose $r_0 \geq R$ in such a way that

$$\begin{aligned}r_0 &\geq \frac{N-1}{\alpha+1}\max\left(\left(\frac{C_1}{\lambda}\right)^{-1/\alpha}, \left(\frac{C_2}{\alpha\lambda}\right)^{-1/(\alpha+1)}\right),\\ u_a(r_0) &> C_1,\\ u_a'(r_0) &> C_2.\end{aligned}$$

Denote

$$K = \max\left\{\left(\frac{C_1}{\lambda}\right)^{-1/\alpha}, \left(\frac{C_2}{\alpha\lambda}\right)^{-1/(\alpha+1)}\right\} + r_0$$

and consider the function $v : [0, K) \to \mathbb{R}^+$,

$$v(r) = \frac{\lambda}{(K-r)^\alpha},$$

which is a solution to the equation

$$v'' - \frac{m^2}{4} v^p = 0, \qquad r \in [0, K).$$

By direct substitution we verify that v satisfies

$$\begin{aligned} &\Delta v \le \frac{m^2}{2} v^p \qquad \text{for } r \in [r_0, K), \\ &v(r_0) \le C_1, \\ &v'(r_0) \le C_2. \end{aligned} \tag{20.19}$$

□

Third step.

LEMMA 20.5. *The inequality*

$$u_a(r) > v(r)$$

holds for all $r \in [r_0, K)$ for which u_a exists.

Proof. Consider the function $\varphi = v - u_a$. By our choice, $\varphi(r_0) < 0$, $\varphi'(r_0) < 0$. If the lemma were not true, a point $R_0 > r_0$ would exist, with $\varphi(R_0) = 0$, $\varphi'(R_0) > 0$, and $v(r) < u_a(r)$ for all $r_0 \le r < R_0$. But (20.18) and (20.19) imply analogously to (19.6) that

$$\varphi'(R_0) R_0^{N-1} \le \frac{m^2}{2} \int_{r_0}^{R_0} (v^p(t) - u_a^p(t)) t^{N-1}\, dt + \varphi'(r_0) r_0^{N-1} < 0,$$

that is $\varphi'(R_0) < 0$, which provides a contradiction.

The lemma is proved, and so is the theorem. □

20.5 A GENERAL THEOREM. NECESSARY AND SUFFICIENT CONDITIONS

There is a gap between conditions imposed on f in the cases considered in Sections 20.3 and 20.4. A more general result may be established. Recall that we consider the problem

$$\begin{aligned} &u'' + \frac{N-1}{r} u' - f(u) = 0, \qquad r \ge 0, \\ &u(0) = a, \\ &u'(0) = 0, \end{aligned} \tag{20.20}$$

where $f : \mathbb{R} \to \mathbb{R}$ is a Lipschitz-continuous function.

This result belongs to Keller [1]:

THEOREM 20.6. *Let $f : \mathbb{R} \to \mathbb{R}$ be a continuous function. Then (20.20) has an entire solution for some a if and only if one of the following statements is valid:*

(1) *there is t_0 such that $f(t_0) = 0$,*
(2) *$f(t) > 0$ for all t and there is t_0 such that $\int_{t_0}^{\infty} F^{-1/2}(t)\,dt = +\infty$,*
(3) *$f(t) < 0$ for all t and there is t_0 such that $\int_{-\infty}^{t_0} F^{-1/2}(t)\,dt = +\infty$,*

where

$$F(t) = \int_{t_0}^{t} f(s)\,ds.$$

Proof. We consider the case $N = 1$. For the general proof see Keller [1].

Case (1) is obvious because we may choose $a = t_0$ and then $u \equiv t_0$.

Now consider case (2). Let $a = t_0$. Then multiplying equation (20.20) by u' and integrating gives

$$\frac{1}{2}|u'(r)|^2 = F(u(r)) > 0 \qquad \text{for } r > 0,$$

because $u'(0) = 0$ and $F'(t) > 0$. Hence,

$$r = \int_{t_0}^{u} (2F(u))^{-1/2} du, \tag{20.21}$$

that is, any large r are available, hence, u_a is an entire solution. Thus, case (2) provides a sufficient condition.

Note that if the integral in case (2) is infinite for some t_0 it diverges also for t_1 instead of t_0. Hence, taking into account (20.21) for arbitrary t_0, we get that the infiniteness condition in (2) is necessary.

Case (3) is analogous to case (2). □

Remark. If f has the form $f = 1 + |t|^p$, $p > 0$, then case (2) of Theorem 20.6 implies $p \le 1$ in correspondence with Theorems 20.3 and 20.4.

§21. Decaying Solutions. The One-dimensional Case

Now we turn to the study of equation (19.2) with the boundary condition

$$u(r) \to 0 \qquad \text{as } r \to \infty. \tag{21.1}$$

DEFINITION. An entire solution satisfying (21.1) is called decaying.

In this section we restrict ourselves to the simplest case $N = 1$, where the questions on existence, nonexistence, and uniqueness receive complete answers. The results in this section are from Berestycki and Lions [1].

21.1 THE MAIN RESULT

Consider the problem

$$\begin{aligned} &u'' - f(u) = 0, \qquad r \geq 0,\\ &u(0) > 0,\\ &u'(0) = 0,\\ &u(r) \to 0 \qquad \text{as } r \to \infty. \end{aligned} \tag{21.2}$$

THEOREM 21.1. *Let $f : \mathbb{R}^+ \to \mathbb{R}$ be a locally Lipschitz function such that*

(i) $f(0) = 0$,

(ii) there exists a finite positive value a such that

$$a = \inf\{t > 0, \quad F(t) = 0\},$$

(iii) $f(a) < 0$.

Then problem (21.2) has a solution with the properties

(a) $u(r) > 0$ *for all* $r \geq 0$,

(b) $u(0) = a$,

(c) $u'(r) < 0$ *for all* $r > 0$.

THEOREM 21.2. *Let problem (21.2) have a solution for some locally Lipschitz function f. Then f satisfies conditions (i), (ii), (iii).*

Remark. A typical function F satisfying the conditions of Theorem 21.1 is $F(t) = |t|^p - |t|^q$ with $2 \leq p < q$.

Remark. Theorems 21.1 and 21.2 provide necessary and sufficient conditions for the solvability of (21.2).

THEOREM 21.3. *A solution of (21.2) under conditions of Theorem 21.1, is unique.*

21.2 THE PROOF OF THEOREM 21.1

Because we claim (*a*), (*b*), (*c*), without loss of generality suppose $F(t) = 0$ for $t \leq 0$ and $F(t) \leq 0$ for $t \geq a$.

Consider equation (21.2) with data (*b*) and $u'(0) = 0$. From Theorem 20.1 there exists an entire solution u_a to (21.2). Identity (20.4) and condition (*ii*) imply

$$\frac{1}{2}|u_a'(R)|^2 - F(u_a(R)) = 0. \tag{21.3}$$

Property (*a*) follows from (21.3) because otherwise there exists a point R_0 with

$$u_a(R_0) = 0, \qquad u_a'(R_0) = 0,$$

which implies in accordance with (*i*) $u_a \equiv 0$ that is impossible by force of (*iii*).

Now show that (*c*) holds. Indeed, if not true, there exists — because $u_a' < 0$ in a neighborhood of the origin due to (*iii*) — a point r_0 such that

$$u_a'(r_0) = 0,$$
$$u_a'(r_0) < 0 \qquad \text{for } 0 < r < r_0,$$

and, by (21.3) and condition (*a*), we have

$$F(u_a(r_0)) = 0,$$
$$0 < u_a(r_0) < a,$$

which contradicts (*ii*).

Finally, (*a*), (*b*), (*c*) imply the existence of a number $0 < b < a$ such that

$$\lim_{r\to\infty} u_a(r) = b$$

and, moreover,

$$\lim_{r\to\infty} u_a'(r) = 0,$$

that is, $F(b) = 0$ by (21.3). By force of (*ii*), we have $b = 0$. □

21.3 The Proof of Theorem 21.2

From (20.4),

$$\frac{1}{2}|u_b'(r)|^2 + F(b) - F(u_b(r)) = 0$$

for any solution u_b with $u(0) = b > 0$. Because $u_b(r) \to 0$ as $r \to \infty$, we have

$$\frac{1}{2}|u_b'(r)|^2 \to -F(b) \qquad \text{as } r \to \infty.$$

Again, because $u_b(r) \to 0$ as $r \to \infty$, the derivative can not converge to a nonzero constant as $r \to \infty$, hence

$$F(b) = 0, \tag{21.4}$$

and thus

$$\frac{1}{2}|u_b'(r)|^2 - F(u_b(r)) = 0 \qquad \text{for all } r \geq 0. \tag{21.5}$$

By (21.5) and the equation, the function u_b at a point r_0, where $F(u_b(r_0)) = 0$, must have a local maximum provided $u''(r_0) = f(u_b(r_0)) < 0$, a local minimum provided $u''(r_0) = f(u_b(r_0)) > 0$, or be constant if $u''(r_0) = f(u_b(r_0)) = 0$ because of the local uniqueness of the Cauchy problem. Hence any nonconstant bounded solution must lie between two neighboring roots of the equation $F(t) = 0$, that is,

$$F(t) \neq 0 \qquad \text{for } 0 < t < b, \tag{21.6}$$

which, together with (21.4), means (*ii*) with $b = a$.

Condition (*ii*) implies the existence of r_0 with $u_b(r_0) = a$. Condition (*iii*) is true because otherwise $u_b(r_0) = a$ at some point r_0 implies $u_b'(r_0) = 0$, $u''(r_0) \geq 0$, that is, $u_b \geq a$, which contradicts decaying.

Condition (*i*) is obvious from equation (21.2) and the decaying of the solution. □

21.4 THE PROOF OF THEOREM 21.3

Let problem (21.2) have a solution u_b with $b > a$. (The case $b < a$ is ruled out because (21.3) implies $F(b) = 0$, but a is minimal among such positive values.) In the previous section, we showed that condition (21.6) holds for any decaying solution. Consequently, we get $F(a) \neq 0$, which contradicts the definition of a. Consequently, we have $a = b$. □

§22. The Phase Plane Method. The Emden-Fowler Equation

In this section, we consider the problem

$$\begin{aligned} &u'' + \frac{N-1}{r}u' + |u|^{p-2}u = 0, \qquad r \geq 0, \\ &u(0) = a, \\ &u'(0) = 0, \end{aligned} \tag{22.1}$$

where $p > 2$. Equation (22.1) is of great physical importance and is known as the Emden-Fowler equation. We present here some results concerning this equation on the basis of the works of Chandrasekhar [1] and Joseph and Lundgren [1].

22.1 SYMMETRY

The Emden-Fowler equation has a symmetry of some kind permitting us to construct all its radial solutions if we know only one.

THEOREM 22.1. *For any $a > 0$ a solution to (22.1) has the form*

$$u_a(r) = a\,v(a^{\frac{p-2}{2}} r)$$

where v is the solution to

$$\begin{aligned} &v'' + \frac{N-1}{r}v' + |v|^{p-2}v = 0, \qquad r \geq 0, \\ &v(0) = 1, \\ &v'(0) = 0. \end{aligned} \tag{22.2}$$

Proof. Existence of an entire solution follows from Theorem 20.1. The fact that u_a satisfies (22.1) follows immediately by substitution. The uniqueness theorem says that there are no other solutions. □

22.2 THE MAIN RESULT

Thus, it is sufficient to study problem (22.2).

Let $N > 2$. Introduce two constants:

$$2^* = \frac{2N}{N-2}, \qquad 2^{**} = \frac{2N-2}{N-2}.$$

Denote

$$\tau = \frac{2}{p-2}.$$

THEOREM 22.2. *Let $2 < p < 2^*$. Then the solution v to (22.2) has infinitely many roots and*

$$\underline{\lim}_{r\to\infty} v(r)\, r^\tau = -\infty, \qquad \overline{\lim}_{r\to\infty} v(r)\, r^\tau = +\infty.$$

Let $p = 2^*$. Then the solution v to (22.2) is positive and

$$\lim_{r\to\infty} v(r)\, r^\tau = 0.$$

Let $p > 2^*$. Then the solution v to (22.2) is positive and

$$\lim_{r\to\infty} v(r)\, r^\tau = (\tau(N-\tau-2))^{\tau/2}.$$

The following sections contain the proof of these statements.

22.3 DECREASING OF THE ORDER

The fact that the original equation has a symmetry suggests that its order may be decreased. Indeed, making the change of variables,

$$z(r) = r^\tau v(r), \qquad r = e^{-t},$$

and choosing t as an independent variable, we get the equation

$$\frac{d^2 z}{dt^2} - (N - 2\tau - 2)\frac{dz}{dt} - \tau(N-\tau-2)z + |z|^{p-2}z = 0. \tag{22.3}$$

Any local solution of (22.3) is defined, by a classical theorem analogous to the theorems of §19, by the data at some point

$$\begin{aligned} z(t_0) &= z_0, \\ \frac{dz}{dt}(t_0) &= y_0. \end{aligned} \tag{22.4}$$

If z and $\frac{dz}{dt}$ remain finite at finite $t \in (-\infty, +\infty)$, then the initial data define an entire solution to (22.3).

Denote $\frac{dz}{dt} = y$. Any solution to (22.3) and (22.4) defines a curve in the plane which may be parametrically presented as

$$\begin{cases} z = z(t), \\ y = y(t). \end{cases}$$

These curves are called the integral curves of equation (22.3). The uniqueness theorem says that an integral curve can not have self-intersection points except for maybe points corresponding to $t = \pm\infty$.

Remark. We pay attention to the fact that whereas r passes from 0 to $+\infty$, t passes from $+\infty$ to $-\infty$. Had we chosen $r = e^t$, we would have obtained a more natural behavior in t and equation (22.3) with changed sign before the second term. The choice made is simply more common for the Emden-Fowler equation.

We may formally choose z as an independent variable and eliminate t,

$$y\frac{dy}{dz} - (N - 2\tau - 2)y - \tau(N - \tau - 2)z + |z|^{p-2}z = 0. \tag{22.5}$$

Analogously, we may choose y as an independent variable and get

$$(|z|^{p-2}z - (N - 2\tau - 2)y - \tau(N - \tau - 2)z)\frac{dz}{dy} + y = 0. \tag{22.6}$$

These transformations are valid only in domains where $\frac{dz}{dy}$ or $\frac{dy}{dz}$, respectively, has constant sign and is finite. In other words, one of equations (22.5) or (22.6) makes sense at the points where the equations

$$\begin{aligned} &y = 0, \\ &|z|^{p-2}z - \tau(N - \tau - 2)z - (N - 2 - 2\tau)y = 0 \end{aligned} \tag{22.7}$$

are incompatible. If a point (z_0, y_0) differs from that defined by (22.7), then it fixes an integral curve to which it belongs. This curve has a tangent defined by the arithmetical formula

$$\frac{dy}{dz}(z_0, y_0) = \frac{(N - 2\tau - 2)y_0 + \tau(N - \tau - 2)z_0 - |z_0|^{p-2}z_0}{y_0} \tag{22.8}$$

or

$$\frac{dz}{dy}(z_0, y_0) = \frac{y_0}{(N - 2\tau - 2)y_0 + \tau(N - \tau - 2)z_0 - |z_0|^{p-2}z_0}. \tag{22.9}$$

Points that are solutions of (22.7) are called singular points. Other points are called regular.

Now consider the boundary condition. We can see that

$$\frac{z(r)}{r^\tau} = v(r) \to 1 \qquad \text{as } r \to 0,$$

$$\frac{y}{r^\tau} = \frac{1}{r^\tau}\frac{dr}{dt}\frac{d}{dr}(r^\tau v(r)) = -(\tau v(r) + rv'(r)) \to -\tau \qquad \text{as } r \to 0.$$

Eliminating r^τ, we have

$$\lim_{z\to 0} y(z) = 0, \qquad \lim_{z\to 0} \frac{y(z)}{z} = -\tau.$$

This means that the initial data define an integral curve leaving the origin and tangent to the line

$$y + \tau z = 0. \tag{22.10}$$

22.4 THE PROOF FOR $2 < p \leq 2^{**}$

Now we start proving Theorem 22.2 directly.

In the case under consideration, equations (22.7) have a unique solution, that is, we have the unique singular point $(0,0)$. We shall study the behavior of the integral curve starting at the origin in accordance with (22.10). For this purpose we draw the curve γ defined by

$$(N - 2 - 2\tau)y + \tau(N - \tau - 2)z - |z|^{p-2}z = 0. \tag{22.11}$$

Because $N - 2\tau - 2 < 0$, $N - \tau - 2 \leq 0$, it has the form depicted in Fig. 22.1.

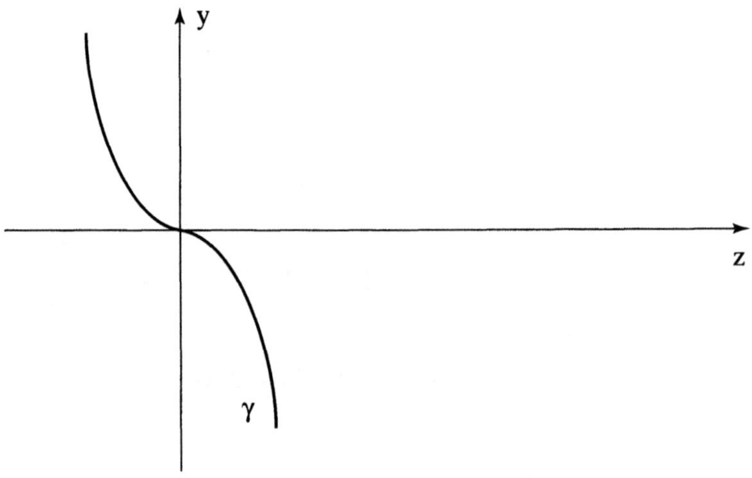

Fig. 22.1

Note that, by (22.8) and (22.9), $\frac{dy}{dz}$ is equal to zero at the regular intersection points of γ and the integral curve σ. Analogously, γ may intersect the line $y = 0$ at regular points only vertically.

We claim that our σ does intersect γ. Indeed, from the equation, we have

$$\frac{dy}{dz} = (N - 2\tau - 2) + \frac{\tau(N - \tau - 2)z - |z|^{p-2}z}{y}. \tag{22.12}$$

Hence, in the open domain (I) (see Fig. 22.2), we have $\frac{dy}{dz} < 0$, and $|\frac{dy}{dz}|$ decreases as z and $|y|$ increase. This means that γ and σ intersect. At the point of intersection, as it was mentioned, $\frac{dy}{dz} = 0$.

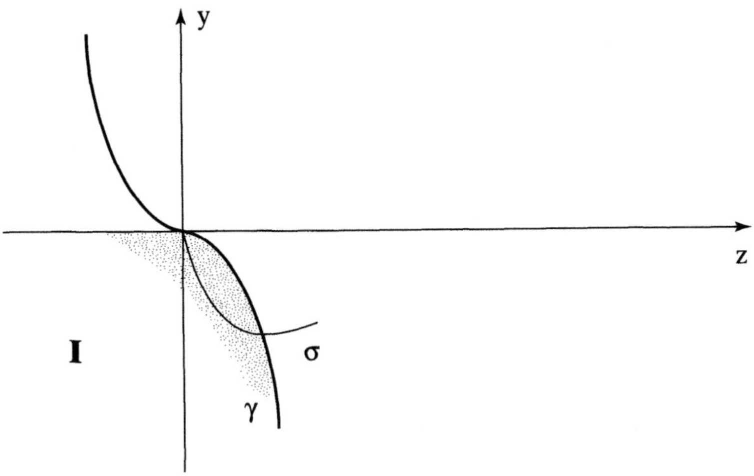

Fig. 22.2

Then, in domain (II) (see Fig. 22.3), we have from (22.11) and (22.12), that $\frac{dy}{dz} > 0$ increases as z and y increase. Hence γ intersects the line $y = 0$, and $\frac{dz}{dy} = 0$ at this point.

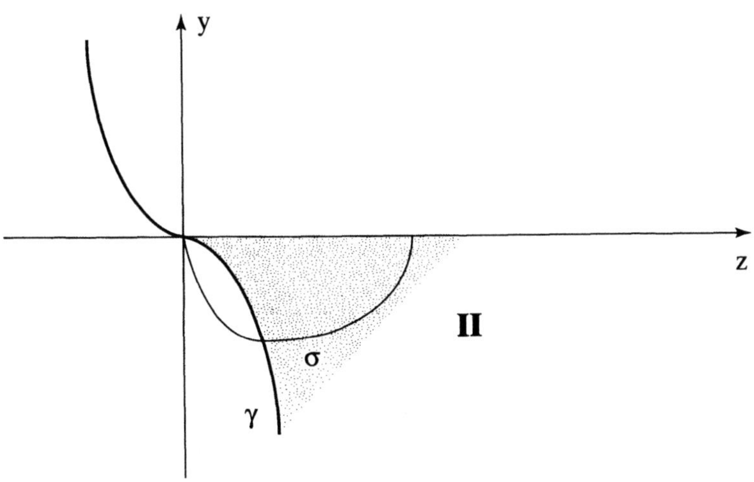

Fig. 22.3

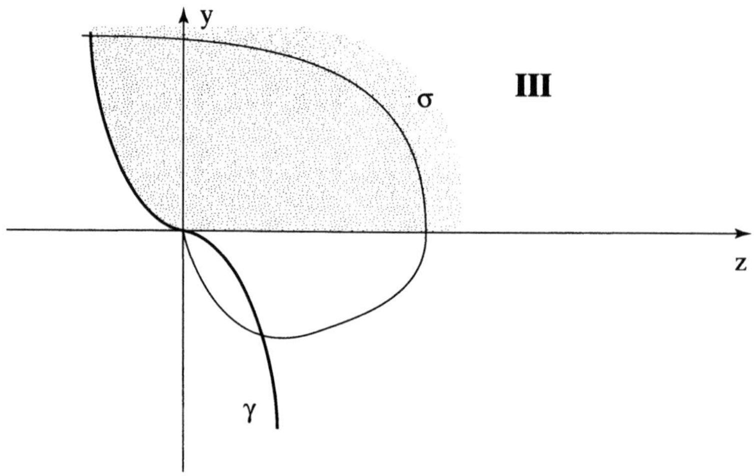

Fig. 22.4

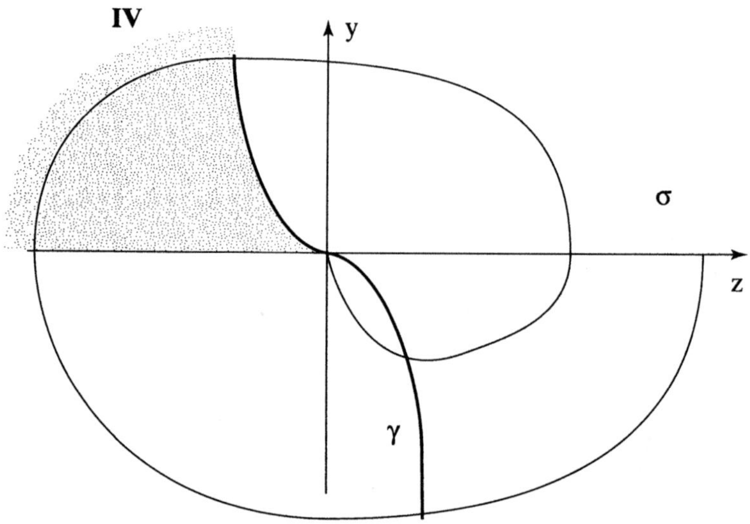

Fig. 22.5

In domain (III) (see Fig. 22.4), we have $\frac{dy}{dz} < 0$, and $|\frac{dy}{dz}|$ decreases as y increases and z decreases. Hence σ again intersects γ at some point at which $\frac{dy}{dz} = 0$.

In domain (IV) (see Fig. 22.5) we have $\frac{dy}{dz} > 0$ and it increases as y and z decrease, and we get an intersection point for σ and the line $y = 0$. The curve σ intersects with domain (I) once more and there $\frac{dy}{dz} < 0$. The Bendixon theorem states:

If functions Z, Y are of the class $C^1(\mathbb{R}^2)$ and the relation

$$\frac{\partial Z}{\partial z}(z,y) + \frac{\partial Y}{\partial y}(z,y)$$

does not change sign and is not identically equal to zero for all $(z,y) \subset \mathbb{R}^2$, then the system

$$\begin{cases} \frac{dz}{dt} = Z(z,y), \\ \frac{dy}{dt} = Y(z,y) \end{cases}$$

does not admit a close integral curve. (The proof may be found in Reissig, Sansone, and Conti [1].)

Equation (22.3) is equivalent to the system

$$\frac{dz}{dt} = y,$$
$$\frac{dy}{dt} = (N - 2\tau - 2)y + \tau(N - \tau - 2)z - |z|^{p-2}z,$$

for which

$$\frac{\partial Z}{\partial z} + \frac{\partial Y}{\partial y} = N - 2\tau - 2 < 0,$$

and it is easy to prove that the curve σ intersects the curve γ once more and then forms a spiral depicted in Fig. 22.5.

Going back to the original variables we see that Theorem 22.2 has been proved for $2 < p \leq 2^{**}$. □

22.5 THE PROOF FOR $2^{**} < p < 2^*$

In this case $N - 2\tau - 2 < 0$, $N - \tau - 2 > 0$, and we have three solutions, $(0,0)$ and $(z_0, y_0) = \pm((\tau(N - \tau - 2))^{\tau/2}, 0)$, to (22.7). Denote the last points as $\theta^{\pm}$. The curve γ has the form shown in Fig. 22.6:

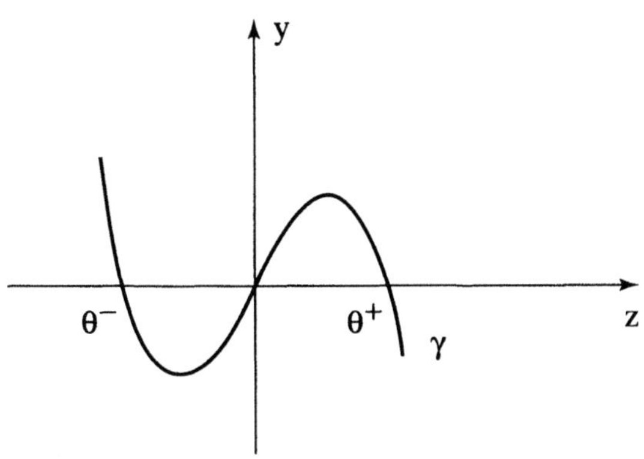

Fig. 22.6

Repeating the reasoning of Section 22.4, we see that σ starting at the origin intersects γ and then the line $y = 0$.

Now σ may intersect γ as shown in Fig. 22.7, left or right.

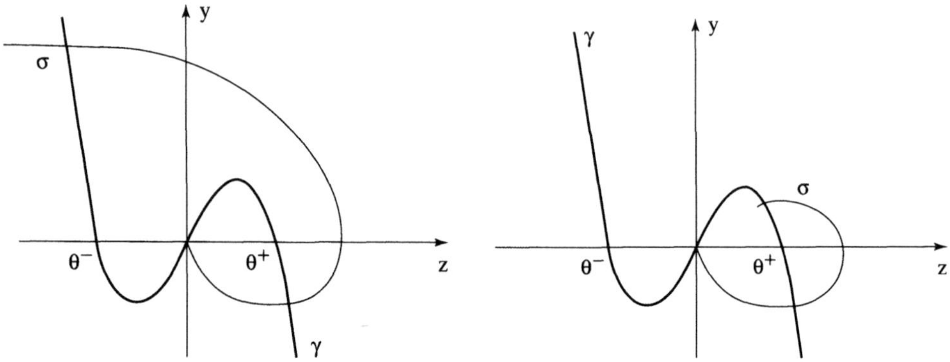

Fig. 22.7

We shall show that the case depicted on the right is impossible. Suppose on the contrary that it is true. Then, taking into consideration that σ may intersect γ only horizontally and $\frac{dy}{dz} > 0$ between σ and γ in the first quadrant, we see that only two variants may take place (see Fig. 22.8).

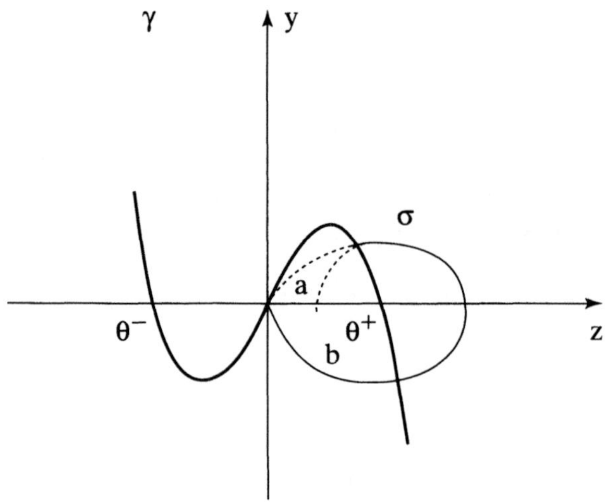

Fig. 22.8

Variant (a) is ruled out by the Bendixson theorem.

We claim now that variant (b) is also impossible. Present the solution in the form

$$z = (\tau(N - \tau - 2))^{\tau/2} + w.$$

Then we have, up to $o(w)$ as $z \to \theta^+$,

$$\frac{d^2w}{dt^2} - (N - 2\tau - 2)\frac{dw}{dt} + 2(N - \tau - 2)w = 0. \tag{22.13}$$

A general solution of this equation, in accordance with the theory of linear ODEs, has the form

$$w = B_1 e^{l_+ t} + B_2 e^{l_- t}$$

where B_1, B_2 are constants and $l_\pm$ are the roots of

$$l^2 - (N - 2\tau - 2)l + 2(N - \tau - 2) = 0$$

equaling

$$l_\pm = \frac{N - 2\tau - 2}{2} \pm \sqrt{\frac{(N - 2\tau - 2)^2}{4} - 2(N - \tau - 2)}. \tag{22.14}$$

Note that $2^{**} < p < 2^*$ implies

$$\operatorname{Re} l_- \le \operatorname{Re} l_+ < 0$$

because $N - 2\tau - 2 < 0$, $N - \tau - 2 > 0$. Now recall the Lyapunov theorem on unstable solutions (see Pontryagin [1]).

THE LYAPUNOV THEOREM. *Let w be a solution to the linear equation*

$$\frac{d^2w}{dt^2} + A\frac{dw}{dt} + Bw = 0, \qquad t \in (-\infty, \infty),$$

where A, B are some constants. Let $\Phi \in C^1(-\infty, +\infty)$ be such that $\Phi(0) = 0$ and $\Phi'(0) = 0$.

If

$$\lim_{t \to -\infty} w(t) = 0 \tag{22.15}$$

implies $w \equiv 0$, then there is a $\delta > 0$ such that if v is a classical solution to

$$\frac{d^2v}{dt^2} + A\frac{dv}{dt} + Bv + \Phi(v) = 0, \qquad t \in (-\infty, 0)$$

with $|v| + |v'| \le \delta$ then v is trivial: $v \equiv 0$.

We see that variant (b) together with the negativeness of $\operatorname{Re} l_\pm$ contradicts the Lyapunov theorem.

Now again, two cases depicted in Fig. 22.9 are possible.

The right-hand case is ruled out analogously to how we rejected the right-hand case of Fig 22.7.

The Bendixon theorem leaves an increasing spiral as its only possible form. □

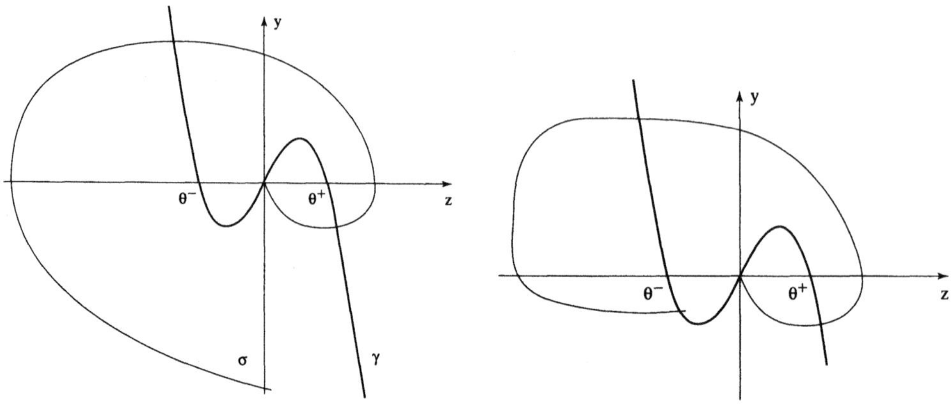

Fig. 22.9

22.6 The Proof for $p = 2^*$

In this case, equation (22.5) has the form

$$y\frac{dy}{dz} - \frac{(N-2)^2}{4}z + |z|^{\frac{4}{N-2}} = 0$$

and can be explicitly integrated. We get in accordance with (22.10)

$$y^2 - \frac{(N-2)^2}{4}|z|^2 + \frac{N-2}{N}|z|^{\frac{2N}{N-2}} = 0.$$

The integral curve depicted in Fig. 22.10 corresponds to this solution in the phase plane.

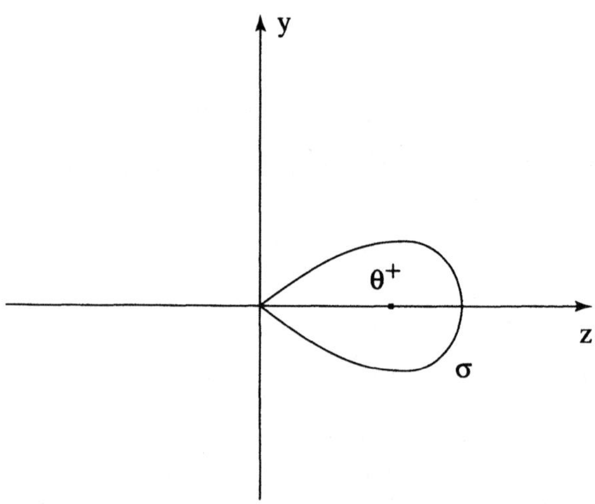

Fig. 22.10

Going back to the original variables, we get

$$v(r) = (1 + (N(N-2))^{-1}r^2)^{\frac{2-N}{2}}. \qquad \square$$

22.7 The Proof for $p > 2^*$

We reason analogously to the preceding sections. From the origin the integral curve intersects the line $y = 0$. From absence Theorem 29.6, it follows that σ can not then intersect the line $z = 0$. Analyzing signs of derivatives as in the previous sections, we see that the integral curve must converge to the point θ^+. The case $p > 2^*$ is thus proved and so is Theorem 22.2. $\square$

22.8 Additional Asymptotics for $p > 2^*$

We want to get some additional information on the behavior of σ in the neighborhood of θ^+.

THEOREM 22.3. *Let $p > 2^*$ and v be a solution of (22.2). Then if*

$$\frac{N-2}{4} < \frac{p-1}{p-2} + \sqrt{\frac{p-1}{p-2}}, \tag{22.16}$$

the equation

$$v(r)r^\tau - (\tau(N-\tau-2))^{\tau/2} = 0$$

has infinitely many roots.

If

$$\frac{N-2}{4} \geq \frac{p-1}{p-2} + \sqrt{\frac{p-1}{p-2}}, \tag{22.17}$$

then

$$v(r)r^\tau < (\tau(N-\tau-2))^{\tau/2}$$

for all $r \geq 0$.

Proof. We proved in the preceding section that σ must be in the neighborhood of θ^+ as $t \to -\infty$ for $p > 2^*$. Let (22.16) be true. Then $l_\pm$ are complex with positive real parts. This means that solutions to linear equation (22.13) behave like decreasing spirals as $t \to -\infty$. The ODE theory says that in this case the solution of our nonlinear equation behaves in the neighborhood of θ^+ in a similar way (we omit the details). The resulting picture is shown in Fig. 22.11.

Now consider the case (22.17) and show that σ intersects γ once at a regular point and then passes to θ^+, that is, it has the form depicted in Fig 22.12.

Moreover, we claim that γ can not intersect the line ξ defined by

$$y = l_+(z - (\tau(N-\tau-2))^{\tau/2}), \tag{22.18}$$

where l_+ is the real positive number defined in (22.14) (see Fig. 22.12).

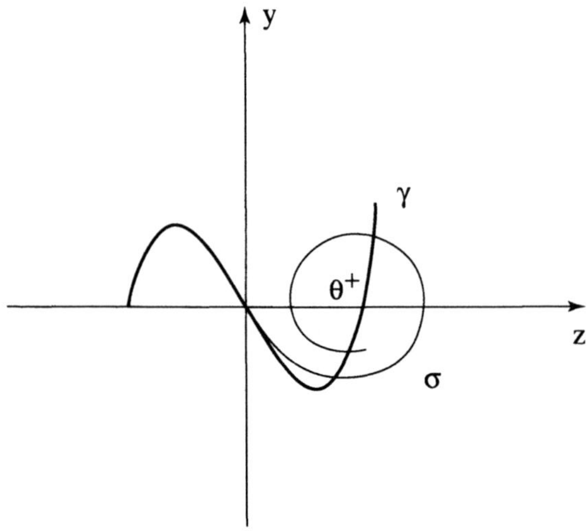

Fig. 22.11

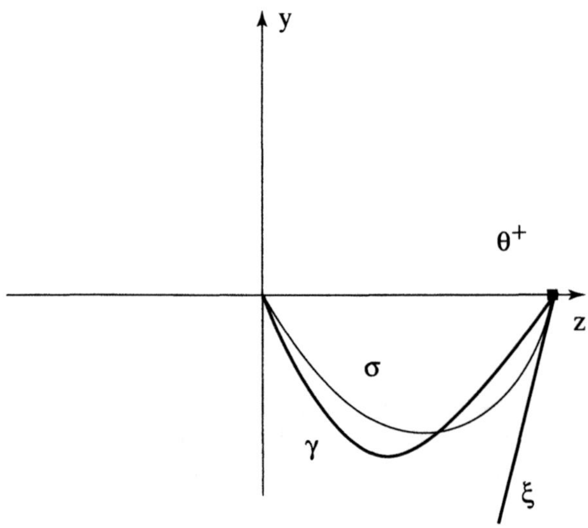

Fig. 22.12

Indeed, substituting (22.18) into (22.12), we see that σ may intersect ξ only with a tangent

$$\frac{dy}{dz} = N - 2\tau - 2 + \frac{\tau(N-\tau-2)z - |z|^{p-2}z}{l_+(z - (\tau(N-\tau-2))^{\tau/2})}.$$

On the other hand, because σ must intersect γ on the left, there must be $\frac{dy}{dz} < l_+$ at a regular intersection point. However,

$$l_+ = N - 2\tau - 2 - \frac{\tau(N-\tau-2)(p-2)}{l_+},$$

therefore we have for $0 < z < \theta_z^+ = (\tau(N-\tau-2))^{\tau/2}$,

$$\begin{aligned}\frac{dy}{dz} - l_+ &= \frac{\tau(N-\tau-2)}{l_+}\left(\frac{z - |z|^{p-2}z/\tau(N-\tau-2)}{z-(\tau(N-\tau-2))^{\tau/2}} + p - 2\right)\\ &= \frac{\tau(N-\tau-2)}{l_+}\left(\frac{w-w^{p-1}}{w-1} + p - 2\right),\end{aligned}$$

where $z = w(\tau(N-\tau-2))^{\tau/2}$.

We see that $\kappa(w) \equiv \frac{w-w^{p-1}}{w-1} + p - 2 > 0$ for all $0 < w < 1$. This follows easily from $\kappa' < 0$ for $0 < w < 1$ (verify!) and

$$\lim_{w \to 1} \kappa(w) = 0.$$

On the other hand, a regular intersection point may lie only in the domain with $0 < w < 1$, and we get a contradiction, which proves the theorem. □

§23. Scaling

Note that investigating the equation

$$-\Delta u + f(u) = 0 \qquad \text{on } \mathbb{R}^N$$

in the radial case, we simultaneously investigated a class of the equations

$$-\Delta u + r^\beta f(u) = 0 \qquad \text{on } \mathbb{R}^n$$

with some β and n. This is a consequence of

THEOREM 23.1. *Let a function u satisfy*

$$u'' + \frac{N-1}{r}u' - f(u) = 0, \qquad r \geq 0.$$

Then v,

$$v(\rho) = u(\rho^{1/\alpha}\alpha), \qquad \alpha = \frac{N-2}{n-2}, \quad n, N > 2,$$

satisfies

$$v'' + \frac{n-1}{\rho}v' - \rho^\beta f(v) = 0, \qquad \rho \geq 0,$$

where

$$\beta = 2\frac{n-N}{N-2}.$$

Proof. It follows immediately by direct substitution. □

§24. Positive Solutions. The Shooting Method

Contrary to the Emden-Fowler equation, the behavior of a solution to the general problem

$$\begin{aligned} &u'' + \frac{N-1}{r}u' - f(u) = 0, \qquad r \geq 0, \\ &u(0) = a > 0, \\ &u'(0) = 0, \end{aligned} \tag{24.1}$$

is more complicated and substantially depends on the initial datum a.

This dependence may be investigated by the so-called shooting method. In this paragraph we are proving the existence of a positive decaying solution with the help of this method.

Recall that the solution to (24.1) with a prescribed a is denoted as u_a.

The main idea of the shooting method is the following. Let a solution u_a fall first on the "ground" $u = 0$ at some point where $u_a' \neq 0$. Let another entire solution u_b with $b > 0$ never fall on the ground (i.e., be positive) but be not decaying. By the theorem on continuous dependence on initial data, we get that there must exist a solution u_c, $c \in (a, b)$, with an intermediate behavior, neither falling, nor flying away off the ground, that is decaying.

The following sections contain the exact formulation. We follow Berestycki, Lions, and Peletier [1].

24.1 THE MAIN RESULT

First, we shall consider an application of the method in an example.

THEOREM 24.1. *Let $N \geq 2$, f be a locally Lipschitz function, and there exist constants $d > b > 0$ such that*

(*i*)

$$\begin{cases} f(t) = 0, & t = 0, \\ f(t) > 0, & 0 < t < b, \\ f(t) = 0, & t = b, \\ f(t) < 0, & t > b, \end{cases} \qquad \begin{cases} F(t) > 0, & 0 < t < d, \\ F(t) = 0, & t = d, \\ F(t) < 0, & d < t < +\infty. \end{cases}$$

(*ii*)

$$\lim_{t \to b+0} \frac{f(t)}{t-b} < 0,$$

(*iii*)

$$\lim_{t \to +\infty} tf(t)/t^q = 0,$$

where $q = 2^*$ if $N > 2$, and $1 < q < +\infty$ is some constant exponent if $N = 2$.

Then there exists such $a > b$ that u_a is a positive decaying solution to (24.1). Moreover,

$$u_a'(r) < 0, \qquad \text{for all } r > 0.$$

Remark. Note that $f'(b) < 0$ is a sufficient condition for (ii).

The rest of this section is devoted to the proof of this fact.

24.2 Classification of Initial Data

Because we are interested in the case $u_a > 0$, suppose without losing generality that $f(t) = -f(-t)$.

We shall search for $a > 0$ such that

$$\begin{cases} u_a(r) > 0 & \text{as } r > 0 \\ u_a'(r) < 0 & \text{as } r > 0, \\ u_a(r) \to 0 & \text{as } r \to \infty, \end{cases}$$

Denote the set of such a as I_0. All we need to prove is that I_0 is not empty.

Consider what other possibilities different from $a \in I_0$ could exist.

Denote by I_+ the subset of $(b, +\infty)$ of such numbers c for which there is a point $R > 0$ with

$$\begin{cases} u_c'(R) = 0, \\ u_c(r) > 0 & \text{as } 0 < r < R. \end{cases}$$

Denote by I_- the subset of $(b, +\infty)$ of such numbers c for which there is a point $R > 0$ with

$$\begin{cases} u_c(R) = 0, \\ u_c'(r) < 0 & \text{as } 0 < r < R. \end{cases}$$

It turns out that other possibilities do not exist:

Lemma 24.2.

$$(b, +\infty) = I_- \cup I_0 \cup I_+.$$

Proof. Theorem 20.1 says that any solution to (24.1) is entire. Obviously, $I_+ \cap I_- = \emptyset$. We shall show that $c \notin I_+ \cup I_-$ implies $c \in I_0$.

Because we consider classical solutions, the derivative u_c' is continuous. Because $f(u_c(0)) < 0$ for $c \in (b, +\infty)$, u_c leaves the origin with a negative derivative.

Let $c \notin I_+ \cup I_-$. Then either

(i) u_c meets the line $x = 0$ with zero derivative, i.e., there exists a point R with

$$u_c(R) = 0, \qquad u_c'(R) = 0$$

implying in turn $u_c \equiv 0$, which is impossible for $c > b$,

(ii) or

$$u_c(r) > 0, \qquad u_c'(r) < 0 \qquad \text{for all } r > 0,$$

which implies

$$\begin{cases} u_c(r) \to l & \text{as } r \to \infty, \\ u_c'(r) \to 0 & \text{as } r \to \infty, \\ u_c''(r) \to 0 & \text{as } r \to \infty \end{cases}$$

for some constant l. From the equation this means that $f(l) = 0$. Then, by (i), we have $l = b$ or $l = 0$.

All we need now is to show that $l \neq b$. Let, on the contrary, $l = b$. Supposing

$$w(r) = (u_c(r) - b)\, r^{\frac{N-1}{2}},$$

we have for sufficiently large r and some positive constant ε,

$$w'' = \left(\frac{f(u_c)}{u_c - b} + \frac{(N-1)(N-3)}{4r^2}\right) w \leq \frac{1}{2}\frac{f(u_c)}{u_c - b} w \leq -\varepsilon\, w < 0. \tag{24.2}$$

This means that w' decreases in r. We see that w' can not tend to a negative value, otherwise $w < 0$, which is impossible because $u \geq b$. Hence, w' converges to a nonnegative value, and w becomes a nondecreasing positive function. Then (24.2) makes w'' negative and bounded away from zero, which contradicts $w' \to$ *a nonnegative value*. Hence, $l = 0$. □

24.3 Nonemptiness of $I_\pm$

Here we claim that I_+ and I_- are nonempty.

Obviously, $(b, d]$ is not empty.

LEMMA 24.3. *$(b, d] \subset I_+$ (which means that I_+ is not empty).*

Proof. If $c \in (b, d]$, then (i) and the definition of d imply $F(c) \geq 0$. On the other hand, by (20.4), $c \in I_-$ implies that at the point $R_c > 0$ where $u_c(R_c) = 0$,

$$F(c) = -\frac{1}{2}|u_c'(R_c)|^2 - (N-1)\int_0^{R_c} \frac{|u_c'(r)|^2}{r}\, dr < 0.$$

If $c \in I_0$, (20.4) implies again that

$$F(c) = -(N-1)\int_0^{\infty} \frac{|u_c'(r)|^2}{r}\, dr < 0.$$

Hence such $c \notin I_- \cup I_0$. Lemma 24.2 then implies $c \in I_+$, which is the desired result. □

Lemma 24.4. *The set I_- is not empty.*

Proof. By the method of §13, it can be proven that the problem

$$-\Delta u + f(u) = 0 \qquad \text{in } B_R,$$
$$u(\partial B_R) = 0,$$

has a positive radial solution under the conditions of Theorem 24.1. It is easy to see that the properties of f guarantee that $u'(r) < 0$ for $0 < r < R$. □

24.4 Openness of $I_\pm$

In order to finish the proof of the theorem we need only to show that the sets $I_\pm$ are open. Then by Lemmas 24.2–24.4 the set I_0 can not be empty.

Lemma 24.5. *The set I_- is open.*

Proof. Let $u_a \in I_-$. Then there exists a point r such that $u_a(r) = 0$ and $u_a'(r) < 0$. This implies that there exists a point $R > r$ such that $u_a(R) < 0$. By the theorem on continuous dependence on data, for all solutions u_c, with c lying in some neighborhood of a, $u(R_c) < 0$ for some R_c. This means that $u(r_c) = 0$ at some other point r_c. Thus, I_- is open. □

Lemma 24.6. *The set I_+ is open.*

Proof. Let $u_a \in I_+$. Then there exists a point r such that $u_a'(r) = 0$, $u(r) > 0$. From the equation, $u_a''(r) \neq 0$, otherwise $u \equiv a$, which is impossible because $b \neq a$. Moreover, $u_a''(r) > 0$ because r is the point of a minimum. Because $u_a \in C^2_{\text{loc}}(\mathbb{R}^N)$, there exist two points $r_1 < r < r_2$ such that $u_a(r) < u_a(r_i)$, $i = 1, 2$. By the theorem on continuous dependence on data, for all solutions u_c, with c lying in some neighborhood of a, $0 < u_c(R) < u_c(R_i)$, $i = 1, 2$, for some points $R_1 < R < R_2$, that is, $u_c'(r_c) = 0$ for some point $R_1 < r_c < R_2$. Thus, I_+ is open. □

Chapter 5
Other Methods

§25. The Method of Upper and Lower Solutions

The Laplace operator possesses a remarkable feature of monotonicity permitting us to study the solvability of elliptic problems. The statement is given in Theorem 25.1.

We consider the equation

$$-\Delta u + f(\mathbf{x}, u) = 0, \tag{25.1}$$

where f is assumed locally Hölder in $\mathbf{x}$ and locally Lipschitz in u.

DEFINITION. A function $u^+ \in C^2(\mathbb{R}^N)$ is called an upper solution to (25.1) if

$$-\Delta u^+ + f(\mathbf{x}, u^+) \geq 0 \qquad \text{on } \mathbb{R}^N.$$

DEFINITION. A function $u^- \in C^2(\mathbb{R}^N)$ is called a lower solution to (25.1) if

$$-\Delta u^- + f(\mathbf{x}, u^-) \leq 0 \qquad \text{on } \mathbb{R}^N.$$

THEOREM 25.1. *Let f satisfy the aforementioned regularity conditions. Assume that there exist upper and lower solutions to equation (25.1) such that*

$$u^+(\mathbf{x}) \geq u^-(\mathbf{x})$$

for all $\mathbf{x} \in \mathbb{R}^N$. Then equation (25.1) has a solution $u \in C^2(\mathbb{R}^N)$ satisfying

$$u^+(\mathbf{x}) \geq u(\mathbf{x}) \geq u^-(\mathbf{x}). \tag{25.2}$$

Proof. See Ako and Kuzano [1] or Ni [1]. □

Choosing different functions from appropriate classes as upper or lower solutions, we get various conditions for the existence of entire solutions.

25.1 Constant Functions

In this section, we choose constant functions as upper or lower solutions.

THEOREM 25.2. *Let there exist constants $C_1 \geq C_2$ such that for all $\mathbf{x} \in \mathbb{R}^N$ the inequalities*

$$f(\mathbf{x}, C_1) \geq 0, \qquad f(\mathbf{x}, C_2) \leq 0$$

hold. Then equation (25.1) has a bounded classical solution u with $C_1 \geq u \geq C_2$.

Proof. Let $u^+ = C_1$ and $u^- = C_2$. It is obvious that

$$-\Delta C_1 + f(\mathbf{x}, C_1) \geq 0,$$
$$-\Delta C_2 + f(\mathbf{x}, C_2) \leq 0,$$

and the result follows from Theorem 25.1. □

EXAMPLE. The equation

$$-\Delta u + |u|^{p-1}u = h(\mathbf{x}) \qquad \text{on } \mathbb{R}^N$$

with $p \geq 1$ and $h \in C^{\alpha}_{\mathrm{loc}}(\mathbb{R}^N)$ has a classical solution $C_1 \geq u \geq C_2$ provided $C_1^p \geq h \geq C_2^p$.

25.2 Radial Test Functions

As usual, we shall be interested later on in decaying solutions, i.e., solutions u with

$$\lim_{|\mathbf{x}| \to \infty} u(\mathbf{x}) = 0.$$

It is easy to see that if $u^{\pm}$ are decaying, then so is u.

Constant functions are simple and give little information. The next step consists in considering radial nonconstant functions.

THEOREM 25.3. *Let $f(\mathbf{x}, u)$ be a function locally Hölder in the first and locally Lipschitz in the second variable. Let functions $y, z \in C^2[0, +\infty)$, with $y \geq z$, satisfy, respectively,*

$$\begin{cases} (r^{N-1}y')' + r^{N-1}f^*(r, y) \leq 0, & r \geq 0, \\ y'(0) = 0, \end{cases}$$

$$\begin{cases} (r^{N-1}z')' + r^{N-1}f_*(r, z) \geq 0, & r \geq 0, \\ y'(0) = 0, \end{cases}$$

where

$$f^*(r, t) = \max_{|\mathbf{x}|=r} f(\mathbf{x}, t),$$
$$f_*(r, t) = \min_{|\mathbf{x}|=r} f(\mathbf{x}, t).$$

Then the equation

$$-\Delta u - f(\mathbf{x}, u) = 0$$

has a classical solution u such that $y \geq u \geq z$.

Proof. Obvious from Theorem 25.1. □

25.3 A SUBLINEAR CASE

Here we consider a first substantial example following Fukagai [1].

$$\begin{aligned} &-\Delta u - a(\mathbf{x})u^p = 0 \qquad \text{on } \mathbb{R}^N, \\ &u \to 0 \qquad \text{as } |\mathbf{x}| \to \infty, \end{aligned} \tag{25.3}$$

with $u > 0$. Here $a : \mathbb{R}^N \to \mathbb{R}$ is a positive locally Hölder function and p is some constant with $0 < p < 1$.

Remark. If we assume in Theorem 25.1 that the corresponding function $f(\mathbf{x}, t)$ in (25.1) is locally Lipschitz in t only on the domain restricted by the maximum of upper and the minimum of lower solutions the statement of this theorem remains true.

Note that the function t^p, $0 < p < 1$, is locally Lipschitz for $t > 0$. Because here we are interested only in positive solutions, this regularity suffices to apply the method of upper and lower solutions if a lower solution is a positive function.

Denote

$$a^*(r) = \max_{|\mathbf{x}|=r} a(\mathbf{x}). \tag{25.4}$$

THEOREM 25.4. *Let $N > 2$ and*

$$\int_0^\infty t a^*(t)\, dt < \infty. \tag{25.5}$$

Then there is a positive decaying solution to (25.3).

Proof. Denote analogously to (25.5)

$$a_*(r) = \min_{|\mathbf{x}|=r} a(\mathbf{x})$$

and define constants $C_1 \geq C_2 \geq 0$ by

$$C_1 = \left(\frac{1}{N-2} \int_0^{+\infty} t\, a^*(t)\, dt \right)^{1/1-p},$$

$$C_2 = \left(\frac{1}{N-2} \left(\int_0^1 t^{N-1}\, a_*(t)\, dt + \int_1^{+\infty} t^{1-p(N-2)}\, a_*(t)\, dt \right) \right)^{1/1-p}.$$

Define now a function $\xi : \mathbb{R}^+ \to \mathbb{R}^+$ by

$$\xi(r) = \begin{cases} C_2, & 0 \le r \le 1, \\ C_2 r^{2-N}, & 1 < r < +\infty, \end{cases}$$

and functions

$$y(r) = \frac{C_1^p}{N-2}\left(\int_0^r \left(\frac{t}{r}\right)^{N-2} t\, a^*(t)\, dt + \int_r^{+\infty} t\, a^*(t)\, dt\right), \tag{25.6}$$

$$z(r) = \frac{C_1^p}{N-2}\left(\int_0^r \left(\frac{t}{r}\right)^{N-2} t\, a_*(t)\xi^p(t)\, dt + \int_r^{+\infty} t\, a_*(t)\xi^p(t)\, dt\right). \tag{25.7}$$

Let us show that y is an upper and z is a lower solution to (25.3). Indeed, it is sufficient by Theorem 25.3 to show that

$$\begin{aligned} &(r^{N-1}y')' + r^{N-1}a^*(r)\, y^p(r) \le 0, \qquad r \ge 0, \\ &y'(0) = 0, \end{aligned} \tag{25.8}$$

$$\begin{aligned} &(r^{N-1}z')' + r^{N-1}a_*(r)\, z^p(r) \ge 0, \qquad r \ge 0, \\ &z'(0) = 0. \end{aligned} \tag{25.9}$$

To verify inequality (25.8) note that by virtue of (25.6)

$$y(r) \le \frac{C_1^p}{N-2}\int_0^{+\infty} t\, a^*(t)\, dt = C_1$$

and as a result the estimate

$$(r^{N-1}y')' = -r^{N-1}a^*(r)\, C_1^p \le r^{N-1}a^*(r) y^p(r)$$

follows.

Analogously, an explicit use of (25.7) and the definition of ξ separately in the case $0 \le r \le 1$ and in the case $1 < r < +\infty$ provide

$$z(r) \ge \xi(r),$$

which in turn implies

$$(r^{N-1}z')' = -r^{N-1}a_*(r)\, \xi^p \ge -r^{N-1}a_*(r) z^p(r)$$

and so proves the inequality (25.9).

Finally, (25.6) and (25.7) imply

$$y(r) \geq z(r) > 0.$$

By virtue of Theorem 25.1 and the remark in this section the existence of a positive classical solution is proved. It remains to show that it is decaying. We claim that

$$y(r) \to 0 \qquad \text{as } r \to \infty.$$

Indeed, we have for the right-hand side of the second term in (25.6), from (25.5),

$$\lim_{r \to \infty} \int_r^{+\infty} t\, a^*(t)\, dt = 0.$$

The first term tends to zero as $r \to +\infty$ because

$$\int_{r^{1/2}}^{r} \left(\frac{t}{r}\right)^{N-2} t\, a^*(t)\, dt \leq \int_{r^{1/2}}^{r} t\, a^*(t)\, dt \leq \int_{r^{1/2}}^{+\infty} t\, a^*(t)\, dt \to 0,$$

$$\int_0^{r^{1/2}} \left(\frac{t}{r}\right)^{N-2} t\, a^*(t)\, dt \leq \frac{1}{r^{N-2}} \int_0^{r^{1/2}} t^{\frac{N-2}{2}} t\, a^*(t)\, dt \leq \frac{\mathbf{C}}{r^{(N-2)/2}} \to 0$$

as $r \to +\infty$. □

25.4 The Anticoercive Case. Positivity Classes

In the following sections we examine the anticoercive superlinear problem

$$-\Delta u - |u|^{p-2}u = h(\mathbf{x}) \qquad \text{in } \mathbb{R}^N, \tag{25.10}$$

where $N > 2$.

We would like to know whether this problem under various conditions on h has a classical solution and, additionally, whether it possesses solutions with some special properties. Knowledge of possible upper and lower solutions would give us the tools for answering these questions. For this purpose consider the operator

$$A_0(u) = -\Delta u - |u|^{p-2}u.$$

It is not generally positive for any $p > 2$, i.e., the condition that the smooth function $u \geq 0$ on $\mathbb{R}^N$ and $u \to 0$ as $|\mathbf{x}| \to \infty$ does not imply the inequality $A_0(u) \geq 0$. Moreover, the inverse to this operator, assuming it exists, is not generally positive, either. This follows, in particular, from the following counterexample.

COUNTEREXAMPLE. Consider the family of functions

$$v_+(\mathbf{x}) = B_+\lambda^{\frac{2}{p-2}}(1+\lambda^2 r^2)^{-\frac{1}{p-2}},$$

with $r = |\mathbf{x}|$, $B_+ > \left(\frac{2N}{p-2}\right)^{\frac{1}{p-2}}$, $\lambda > 0$, and $p > 2$. Then for this family of functions we have $v_+ > 0$ and $v_+(r) \to 0$ as $r \to \infty$, and $A_0(v_+(r)) < 0$, $A_0(v_+(r)) \to 0$ as $r \to \infty$. Correspondingly, for the family of functions

$$v_-(\mathbf{x}) = B_-\lambda^{\frac{2}{p-2}}(1+\lambda^2 r^2)^{-\frac{1}{p-2}},$$

with $B_- < -\left(\frac{2N}{p-2}\right)^{\frac{1}{p-2}}$, $\lambda > 0$, and $p > 2$, we have $v_- < 0$ and $v_-(r) \to 0$ as $r \to \infty$, and $A_0(v_-(r)) > 0$, $A_0(v_-(r)) \to 0$ as $r \to \infty$.

Nevertheless, it turns out that for appropriate $p > 2$ there exist classes of smooth functions u for which the condition $u \geq 0$ implies $A_0(u) \geq 0$.

To show this consider, for $p > 2$ and $N > 2$, the family of functions

$$u_+(\mathbf{x}) = C\,\lambda^{\frac{2}{p-2}}(1+\lambda^2 r^2)^{\beta} \tag{25.11}$$

depending on the real parameters $C > 0$, $\lambda > 0$, and β.

For this family we have

$$\begin{aligned} A_0(u) = &-2\beta\lambda^{2\frac{p-1}{p-2}}C\,(N+2\beta-2)\,(1+\lambda^2 r^2)^{\beta-1} \\ &+ 4\beta(\beta-1)\lambda^{2\frac{p-1}{p-2}}C\,(1+\lambda^2 r^2)^{\beta-2} - \lambda^{2\frac{p-1}{p-2}}C^{p-1}(1+\lambda^2 r^2)^{\beta(p-1)}. \end{aligned}$$

Then the positivity condition for the operator A_0 on this family assumes the form

$$-2\beta(N+2\beta-2) + 4\beta(\beta-1)\frac{1}{1+\lambda^2 r^2} - C^{p-2}(1+\lambda^2 r^2)^{\beta(p-2)+1} \geq 0 \tag{25.12}$$

for any $r \geq 0$.

For this relation to hold, it is obviously necessary that

$$\beta \leq \beta_3 = -\frac{1}{p-2},$$

and for $\beta < \beta_3$, it is necessary that

$$\beta \geq \beta_1 = -\frac{N-2}{2}.$$

Note that the inequality $\beta_1 \leq \beta < \beta_3$ implies the following condition for p:

$$p > 2^{**} = \frac{2N-2}{N-2}, \qquad N > 2.$$

It is clear that, provided $N \geq 3$, $p > 2^{**}$, for any $\beta : \beta_1 < \beta \leq \beta_3$ there exists a constant $C(\beta, p, N)$ such that the operator A_0 on functions of the form (25.11) with $C : 0 < C < C(\beta, p, N)$ is positive (nonnegative).

Below, we shall study three such classes of functions with different asymptotics as $|\mathbf{x}| \to \infty$, corresponding to the values $\beta = \beta_1$ for $p \geq 2^*$, $\beta = \beta_2 = -\frac{2}{p-2} \geq \beta_1$ for $p \geq 2^*$, and $\beta = \beta_3$ for $p > 2^{*}*$.

To begin with, consider the class of functions corresponding to the smallest value $\beta = \beta_1$.

The first class of functions for $p \geq 2^*$

In this case we have

$$u_1(\mathbf{x}) = C_1\lambda^{\frac{2}{p-2}}(1+\lambda^2 r^2)^{\beta_1}, \tag{25.13}$$

with $\beta_1 = -\frac{N-2}{2}$, $C_1 > 0$, and $\lambda > 0$.

Positivity condition (25.12) for the operator A_0 on this first class of functions assumes the form

$$N(N-2) - C_1^{p-2}(1+\lambda^2 r^2)^{-\frac{(N-2)p-2N}{2}} \geq 0 \qquad \text{for all } r \geq 0.$$

Hence it necessarily follows that

$$p \geq 2^* = \frac{2N}{N-2},$$
$$0 < C_1 \leq C_1^* = [(N-2)N]^{\frac{1}{p-2}}. \tag{25.14}$$

Clearly, the conditions are also sufficient.

Thus, the first class of functions on which A_0 is positive exists for $N \geq 3$, $p \geq 2^*$, and is defined by (25.13) with $\beta_1 = -\frac{N-2}{2}$, $\lambda > 0$, and C_1 satisfying inequality (25.14).

Let us now calculate the right-hand side of the operator A_0 on this class of functions. We have

$$h_1 \equiv A_0(u_1) = \frac{C_1\lambda^{2\frac{p-1}{p-2}}}{(1+\lambda^2 r^2)^{\frac{N+2}{2}}}\left[(N-2)N - \frac{C_1^{p-2}}{(1+\lambda^2 r^2)^{\frac{N-2}{2}(p-2^*)}}\right]. \tag{25.15}$$

Note that for $p = 2^*$ and $C_1 = C_1^* = [(N-2)N]^{1/(p-2)}$ we get the class of functions

$$u_{1*}(\mathbf{x}) = \frac{[N(N-2)]^{(N-2)/4}\lambda^{(N-2)/2}}{(1+\lambda^2 r^2)^{(N-2)/2}},$$

corresponding to the critical exponent $p = 2^*$ on which $A_0(u_{1*}(\mathbf{x})) \equiv 0$.

Note that Gidas, Ni, and Nirenberg [1] proved that any positive classical solution of the equation

$$\Delta u + |u|^{2^*-2}u = 0$$

for $N \geq 3$ with finite energy

$$\int_{\mathbb{R}^N} |\nabla u|^2 \, d\mathbf{x} < +\infty$$

necessarily has the form

$$u(\mathbf{x}) = u_{1*}(\mathbf{x}-\mathbf{x}_0) = \frac{[N(N-2)]^{(N-2)/4}\lambda^{(N-2)/2}}{(1+\lambda^2|\mathbf{x}-\mathbf{x}_0|^2)^{(N-2)/2}}, \qquad \mathbf{x}_0 \in \mathbb{R}^N.$$

Before proceeding to the next class of functions corresponding to the other boundary value $\beta = \beta_3 = -1/(p-2)$, consider the intermediate class of functions corresponding to $\beta = \beta_2 = -2/(p-2)$.

The second class of functions for $p \geq 2^*$

In this case we have

$$u_2(\mathbf{x}) = C_2 \lambda^{\frac{2}{p-2}} (1 + \lambda^2 r^2)^{\beta_2}, \tag{25.16}$$

with $\beta_2 = -\frac{2}{p-2}$, $C_2 > 0$, and $\lambda > 0$.

Positivity condition (25.12) for the operator A_0 on this class of functions assumes the form

$$4\frac{(N-2)(p-2^*)}{(p-2)^2} + \frac{8p(p-2)^{-2} - C_2^{p-2}}{1 + \lambda^2 r^2} \geq 0 \qquad \text{for all } r > 0.$$

Hence it necessarily follows that

$$p \geq 2^*$$

and

$$0 < C_2 \leq C_2^* = \left[4\frac{(N-2)(p-2^*) + 2p}{(p-2)^2}\right]^{\frac{1}{p-2}}. \tag{25.17}$$

Clearly, the conditions are also sufficient.

Thus, the second class of functions on which the operator A_0 is positive exists for $N \geq 3$, $p \geq 2^*$, and is defined by (25.16) with $\beta_2 = -\frac{2}{p-2}$, $\lambda > 0$, and C_2 satisfying inequality (25.17).

Note that $\beta_1 = \beta_2$ for $p = 2^*$, and in this case both classes of functions coincide. In the case $p > 2^*$ we have $\beta_2 > \beta_1$ and these two classes are different with different asymptotics as $|\mathbf{x}| \to \infty$.

Let us now calculate the right-hand side of the operator A_0 on the second class of functions. We have

$$h_2 \equiv A_0(u_2) = \frac{C_2 \lambda^{2\frac{p-1}{p-2}}}{(1 + \lambda^2 r^2)^{\frac{p}{p-2}}} \left[4\frac{(N-2)(p-2^*)}{(p-2)^2} + \frac{8p(p-2)^{-2} - C_2^{p-2}}{1 + \lambda^2 r^2}\right]. \tag{25.18}$$

Consider now the next class of functions, corresponding to the boundary value $\beta = \beta_3 = -\frac{1}{p-2}$, on which the operator A_0 is positive.

The third class of functions for $p > 2^{**} = \frac{2N-2}{N-2}$

In this case we have

$$u_3(\mathbf{x}) = C_3 \lambda^{\frac{2}{p-2}} (1 + \lambda^2 r^2)^{\beta_3}, \tag{25.19}$$

with $\beta_3 = -\frac{1}{p-2}$, $C_3 > 0$, and $\lambda > 0$. Then

$$h_3 \equiv A_0(u_3)$$
$$= \frac{C_3 \lambda^{2\frac{p-1}{p-2}}}{(1 + \lambda^2 r^2)^{\frac{p-1}{p-2}}} \left[2\frac{(N-2)(p-2^*)}{(p-2)^2} - C_3^{p-2} + 4\frac{p-1}{(p-2)^2}\frac{1}{(1 + \lambda^2 r^2)}\right]. \tag{25.20}$$

Obviously, the positivity condition for the operator A_0 on this class of functions is satisfied if

$$p > 2^* = \frac{2N-2}{N-2} \qquad (N \geq 3)$$

and

$$0 < C_3 \leq C_3^* = \left[\frac{2(N-2)(p-2^*)}{(p-2)^2}\right]^{\frac{1}{p-2}}. \tag{25.21}$$

Remark. From (25.20) it follows that, for $p > 2^*$ and $N \geq 3$, the operator A_0 on the class of functions

$$u(\mathbf{x}) = C\,\lambda^{\frac{2}{p-2}}(1+\lambda^2 r^2)^{-\frac{1}{p-2}}$$

with $\lambda > 0$ and $C > 0$ is positive if

$$0 < C \leq C_3^*,$$

it ceases being positive if $C > C_3^*$, and it becomes "antipositive", i.e., $u(\mathbf{x}) \geq 0$ implies $A(u(\mathbf{x})) \leq 0, \mathbf{x} \in \mathbb{R}^N$ if

$$C \geq \left(\frac{2N}{p-2}\right)^{\frac{1}{p-2}}.$$

25.5 Solvability of Anticoercive Problems in the Classes of Functions with Prescribed Asymptotics

We shall apply the results of the previous section to problem (25.1) as regards its solvability in the classes of functions with prescribed asymptotics as $|\mathbf{x}| \to \infty$, choosing functions from corresponding positivity classes of the operator A_0 as upper or lower solutions. We shall be considering appropriate right-hand sides h.

We represent f in the form

$$f(\mathbf{x},t) = -|t|^{p-2}t + f_1(\mathbf{x},t) - h(\mathbf{x})$$

Then problem (25.1) assumes the form

$$\begin{aligned} &-\Delta u - |u|^{p-2}u + f_1(\mathbf{x},u) = h(\mathbf{x}) \quad \text{in } \mathbb{R}^N, \\ &u \to 0 \qquad \text{as } |\mathbf{x}| \to \infty, \end{aligned} \tag{25.22}$$

Throughout this section, we assume that $f_1(\mathbf{x},t)$ is locally Hölder in $\mathbf{x}$ and locally Lipschitz in t, and h is locally Hölder.

Solvability in the first class

This class of functions $u \in C^2_{\text{loc}}(\mathbb{R}^N)$ is defined by the following condition: for each function u there exist constants $C > 0$ and $\lambda > 0$ such that

$$|u(\mathbf{x})| \leq \frac{C}{(1+\lambda^2|\mathbf{x}|^2)^{\frac{N-2}{2}}} \qquad x \in \mathbb{R}^N, \qquad N \geq 3.$$

THEOREM 25.4. *Let there exist $\lambda > 0$ and $C > 0$ such that*

$$f_1(\mathbf{x}, -u_1(\mathbf{x})) + h_1(\mathbf{x}) \leq h(\mathbf{x}) \leq f_1(\mathbf{x}, u_1(\mathbf{x})) - h_1(\mathbf{x}) \qquad \mathbf{x} \in \mathbb{R}^N,$$

where the functions u_1 and h_1 are defined by (25.13) and (25.15), respectively, with $C_1 = C_1^*$ from (25.14).

Then, for $N \geq 3$, $p > 2^*$, problem (25.22) has a solution $u \in C^2_{\text{loc}}(\mathbb{R}^N)$ satisfying the inequality

$$|u(\mathbf{x})| \leq u_1(\mathbf{x}).$$

Proof. It follows immediately from Theorem 25.1 where $\pm u_1$ serve as upper or lower solutions, respectively. □

Positive solutions in the first class

Everywhere below, the term positive solutions will be understood as meaning nonnegative solutions.

THEOREM 25.5. *Let there exist $\lambda > 0$ and $C > 0$ such that*

$$f_1(\mathbf{x}, 0) \leq h(\mathbf{x}) \leq f_1(\mathbf{x}, u_1(\mathbf{x})) - h_1(\mathbf{x})$$

with functions u_1 and h_1 as in the previous theorem.

Then, for $N \geq 3$ and $p > 2^*$, problem (25.22) has a positive (nonnegative) solution $u \in C^2_{\text{loc}}(\mathbb{R}^N)$ satisfying the inequality

$$0 \leq u(\mathbf{x}) \leq u_1(\mathbf{x}).$$

Proof. It follows from Theorem 25.1 where u_1 and 0 serve as upper or lower solutions, respectively. □

Solvability in the second class

This class of functions $u \in C^2_{\text{loc}}(\mathbb{R}^N)$ is defined by the following condition: for each function u there exist constants $C > 0$ and $\lambda > 0$ such that

$$|u(\mathbf{x})| \leq \frac{C\lambda^{\frac{2}{p-2}}}{(1+\lambda^2|\mathbf{x}|^2)^{\frac{2}{p-2}}} \qquad x \in \mathbb{R}^N, \qquad N \geq 3.$$

THEOREM 25.6. *Let there exist $\lambda > 0$ and $C > 0$ such that*

$$f_1(\mathbf{x}, -u_2(\mathbf{x})) + h_2(\mathbf{x}) \leq h(\mathbf{x}) \leq f_1(\mathbf{x}, u_2(\mathbf{x})) - h_2(\mathbf{x}) \qquad \mathbf{x} \in \mathbb{R}^N,$$

where the functions u_2 and h_2 are defined by (25.16) and (25.18), respectively, with $C_2 = C_2^*$ from (25.17).

Then, for $N \geq 3$ and $p > 2^*$, problem (25.22) has a solution $u \in C^2_{\mathrm{loc}}(\mathbb{R}^N)$ satisfying the inequality

$$|u(\mathbf{x})| \leq u_2(\mathbf{x}), \qquad \mathbf{x} \in \mathbb{R}^N.$$

Proof. It follows from Theorem 25.1 where $\pm u_2$ serve as upper or lower solutions, respectively. □

Positive solutions in the second class

THEOREM 25.7. *Let there exist $\lambda > 0$ and $C > 0$ such that*

$$f_1(\mathbf{x}, 0) \leq h(\mathbf{x}) \leq f_1(\mathbf{x}, u_2(\mathbf{x})) - h_2(\mathbf{x}), \qquad \mathbf{x} \in \mathbb{R}^N,$$

with functions u_2 and h_2 as in the previous theorem.

Then, for $N \geq 3$ and $p > 2^*$, problem (25.22) has a positive (nonnegative) solution $u \in C^2_{\mathrm{loc}}(\mathbb{R}^N)$ satisfying the inequality

$$0 \leq u(\mathbf{x}) \leq u_2(\mathbf{x}), \qquad \mathbf{x} \in \mathbb{R}^N. \tag{25.23}$$

Proof. It follows from Theorem 25.1 where u_2 and 0 serve as upper or lower solutions, respectively. □

EXAMPLE. Consider the problem

$$\begin{aligned} &-\Delta u + \mu u - |u|^{p-2}u = h(\mathbf{x}) \qquad \text{on } \mathbb{R}^N, \\ &u \to 0 \qquad \text{as } |\mathbf{x}| \to \infty, \end{aligned} \tag{25.24}$$

under the following conditions:

Let $\mu \geq 0$, $N \geq 3$, $p > 2^*$, $h \in C^{\nu}_{\mathrm{loc}}(\mathbb{R}^N)$ with some $\nu \in (0,1)$, and let there exist $\lambda > 0$ such that

$$0 \leq h(\mathbf{x}) \leq \left[4\frac{(N-2)(p-2^*)}{(p-2)^2}\frac{\lambda^2}{1+\lambda^2|\mathbf{x}|^2} + \mu\right] \left[\frac{8p}{(p-2)^2}\right]^{\frac{1}{p-2}} \lambda^{\frac{2}{p-2}}(1+\lambda^2|\mathbf{x}|^2)^{-\frac{2}{p-2}}, \qquad \mathbf{x} \in \mathbb{R}^N.$$

Then there exists a nonnegative classical solution to problem (25.24) satisfying (25.23).

Solvability in the third class

This class of functions $u \in C^2_{\mathrm{loc}}(\mathbb{R}^N)$ is defined by the following condition: for each function u there exist constants $C > 0$ and $\lambda > 0$ such that

$$|u(\mathbf{x})| \leq \frac{C\lambda^{\frac{2}{p-2}}}{(1+\lambda^2|\mathbf{x}|^2)^{\frac{1}{p-2}}} \qquad x \in \mathbb{R}^N, \qquad N \geq 3.$$

THEOREM 25.8. *Let there exist $\lambda > 0$ and $C > 0$ such that*

$$f_1(\mathbf{x}, -u_3(\mathbf{x})) + h_3(\mathbf{x}) \leq h(\mathbf{x}) \leq f_1(\mathbf{x}, u_3(\mathbf{x})) - h_3(\mathbf{x}), \qquad \mathbf{x} \in \mathbb{R}^N,$$

where the functions u_3 and h_3 are defined by (25.19) and (25.20), respectively, with $C_3 = C_3^*$ from (25.21).

Then, for $N \geq 3$ and $p > 2^*$, problem (25.22) has a classical solution satisfying the inequality

$$|u(\mathbf{x})| \leq u_3(\mathbf{x}).$$

Proof. It follows from Theorem 25.1 where $\pm u_3$ serve as upper or lower solutions, respectively. □

Positive solutions in the third class

THEOREM 25.9. *Let there exist $\lambda > 0$ and $C > 0$ such that*

$$f_1(\mathbf{x}, 0) \leq h(\mathbf{x}) \leq f_1(\mathbf{x}, u_3(\mathbf{x})) - h_3(\mathbf{x}), \qquad \mathbf{x} \in \mathbb{R}^N,$$

with the functions u_3 and h_3 as in the previous theorem.

Then, for $N \geq 3$ and $p > 2^*$, problem (25.22) has a positive (nonnegative) solution $u \in C^2_{\mathrm{loc}}(\mathbb{R}^N)$ satisfying the inequality

$$0 \leq u(\mathbf{x}) \leq u_3(\mathbf{x}), \qquad \mathbf{x} \in \mathbb{R}^N.$$

Proof. It follows from Theorem 25.1 where u_3 and 0 serve as upper or lower solutions, respectively. □

EXAMPLE. Consider problem (25.24) under the following conditions:

Let $\mu \geq 0$, $N \geq 3$, $p > 2^*$, $h \in C^{\nu}_{\mathrm{loc}}(\mathbb{R}^N)$ with some $\nu \in (0,1)$, and $\lambda > 0$ such that

$$0 \leq h(\mathbf{x}) \leq \left[4\frac{p-1}{(p-2)^2}\frac{\lambda^2}{1+\lambda^2|\mathbf{x}|^2} + \mu\right] \left[\frac{2((N-2)p-2N+2)}{(p-2)^2}\right]^{\frac{1}{p-2}} \lambda^{\frac{2}{p-2}}(1+\lambda^2|\mathbf{x}|^2)^{-\frac{1}{p-2}}, \qquad \mathbf{x} \in \mathbb{R}^N.$$

Then there exists a nonnegative classical solution u to problem (25.24) satisfying (25.23).

Remark. In this section we used the work of Pohozaev [4].

Many other applications of the method of upper and lower solutions are contained, for instance, in the works of Furusho [1] and Furusho and Kuzano [1]. See also references therein.

§26. The Leray-Schauder Method

In this section we demonstrate another approach to the existence question for coercive problems based on the Leray-Schauder method. The method is based on an observation known from nonlinear functional analysis as the Leray-Schauder fixed point principle.

We introduce the notion of a weakly continuous operator. Let X be a Banach space.

Definition. An operator $A : X \to X$ is called weakly continuous if it transforms any weakly convergent sequence into a strongly convergent one.

For such operators the following fixed point theorem is true:

Schauder Theorem. *Let X be a reflexive Banach space and D_R be a ball of radius R in X. Let an operator A be weakly continuous on X and map the ball D_R into itself. Then there exists an element $u \in X$ such that*

$$u = A(u) \qquad \text{in } X.$$

Proof. See Krasnoselskii [1] and Vainberg [1]. □

We show in the following sections how this topological result can be applied to nonlinear elliptic problems in $\mathbb{R}^N$.

26.1 The Leray-Schauder Theorem

The direct application of the Schauder theorem to elliptic equations is not convenient. A more pertinent variant is the following:

Leray-Schauder Theorem. *Let X be a reflexive Banach space. Let an operator $A : X \to X$ be weakly continuous on X. Assume that there exists a constant $R > 0$ such that for any possible solution $(\lambda, u) \in [0, 1] \times X$ to the problem*

$$u = \lambda A(u) \qquad \text{in } X \tag{26.1}$$

the inequality

$$\|u\|_X < R \tag{26.2}$$

holds.

Then the equation

$$u = A(u) \qquad \text{in } X \tag{26.3}$$

has a solution $u \in X$.

Remark. We draw attention to the fact that the assumptions of the theorem do not require that problem (26.1) indeed has a solution. We require only that any possible solution to (26.1) implies estimate (26.2). If that is true, we may prove the real existence of a solution to (26.3).

26.2 Proof of the Leray-Schauder Theorem

Consider the ball D_R in the space X centered at the origin, with R from (26.2). Therefore,

$$u \neq \lambda A(u) \text{ for all } \lambda \in [0,1] \text{ and } u \in X \text{ with } \|u\|_X = R. \tag{26.4}$$

Consider the operator $A_1 : D_R \to D_R$, defined as

$$A_1(u) = \begin{cases} A(u), & \|A(u)\|_X \leq R, \\ R\,A(u)\,\|A(u)\|_X^{-1}, & \|A(u)\|_X > R. \end{cases}$$

This operator is weakly continuous and maps the ball D_R of the reflexive space X into itself. Consequently, the Schauder theorem implies the existence of an element $u_0 \in D_R$ such that

$$u_0 = A_1(u_0) \tag{26.5}$$

that is, $A_1(u_0) = A(u_0)$.

We now show that u_0 is a solution to (26.3) by way of contradiction. If u_0 is a solution to (26.5) but not a solution to (26.3), we have

$$\|A(u_0)\|_X > R$$

and

$$u_0 = \frac{R\,A(u_0)}{\|A(u_0)\|_X}.$$

In other words,

$$u_0 = \lambda\, A(u_0) \qquad \text{with } \lambda = \frac{R}{\|A(u_0)\|_X} < 1,$$

which contradicts (26.4). □

26.3 An Example

Here we demonstrate the Leray-Schauder method on a simple example.

Consider the problem

$$\begin{aligned} -\Delta u + g(\mathbf{x}, u) = h(\mathbf{x}) \qquad & \text{in } \mathbb{R}^N, \\ u \to 0 \qquad \text{as } |\mathbf{x}| \to \infty, & \end{aligned} \tag{26.6}$$

where g is a Carathéodory function and h is measurable. Recall that $\mathcal{E}_{2^*}$ was defined in Section 1.1.

THEOREM 26.1. *Let $N > 2$, $h \in \mathcal{E}_{2^*}^*$, and let there be exponents $1 < p < 2^*$ and γ,*

$$\gamma < \frac{p(N-2)-2N}{4}, \tag{26.7}$$

such that for almost all $\mathbf{x} \in \mathbb{R}^N$ and all $t \in \mathbb{R}$

$$|g(\mathbf{x},t)| \le \mathbf{C}|t|^{p-1}(1+|\mathbf{x}|^2)^\gamma.$$

Let also

$$t\, g(\mathbf{x},t) \ge 0.$$

Then problem (26.6) has a solution $u \in \mathcal{E}_{2^*}$.

Before proving the theorem we establish a preliminary result.

LEMMA 26.2. *Under the above conditions the mapping $g(\mathbf{x}, .) : \mathcal{E}_{2^*} \to \mathcal{E}_{2^*}^*$ is weakly continuous.*

Proof. Let $u_n \rightharpoonup u$ in $\mathcal{E}_{2^*}$. By the Hölder inequality, it is easy to see that for all $v \in \mathcal{E}_2$,

$$|\langle g(u_n)-g(u),v\rangle| \le \|v\|_{2^*} \left(\int\limits_{\mathbb{R}^N} |g(\mathbf{x},u_n)-g(\mathbf{x},u)|^{2'}\, d\mathbf{x} \right)^{1/2'},$$

where we denote $2' = 2^*/(2^*-1)$. By the Hölder inequality and (26.7), we have

$$\int\limits_{\mathbb{R}^N\setminus B_R} |g(\mathbf{x},u_n)-g(\mathbf{x},u)|^{2'}\, d\mathbf{x}$$

$$\le \left(\int\limits_{\mathbb{R}^N\setminus B_R} |g(\mathbf{x},u_n)-g(\mathbf{x},u)|^{2't}(1+|\mathbf{x}|^2)^{-2'\gamma t}\, d\mathbf{x} \right)^{1/t}$$

$$\times \left(\int\limits_{\mathbb{R}^N\setminus B_R} (1+|\mathbf{x}|^2)^{2'\gamma t'}\, d\mathbf{x} \right)^{1/t'} \le C(R)\left(\|u\|_{2^*}^{p-1} + \|u_n\|_{2^*}^{p-1}\right), \tag{26.8}$$

where $t = \frac{N+2}{(p-1)(N-2)}$, $t' = t/(t-1)$, and hence, for any $\varepsilon > 0$ there is an R_ε such that the right-hand side of (26.8) is less than ε. By the Kondrashov theorem, given ε and R_ε, we may find n_ε such that for all $n \ge n_\varepsilon$

$$\int\limits_{B_R} |g(\mathbf{x},u_n)-g(\mathbf{x},u)|^{2'}\, d\mathbf{x} < \varepsilon.$$

Hence,

$$|\langle g(u)-g(u_n),v\rangle| \le \mathbf{C}\,\varepsilon\,\|v\|_{2^*},$$

where $\mathbf{C}$ does not depend on v and ε. Letting $\varepsilon \to 0$ and taking into consideration the definition of a dual space we get the result. □

Proof of Theorem 26.1. The problem (26.6) is equivalent to the problem

$$u = -\Delta^{-1}(h(\mathbf{x}) - g(\mathbf{x},u)), \qquad u \in \mathcal{E}_{2^*},$$

where Δ^{-1} is the inverse to Δ defined by the Riesz theorem. By Lemma 26.2, $A = -\Delta^{-1}h + \Delta^{-1}g(\mathbf{x},\,.\,)$ is weakly continuous from $\mathcal{E}_{2^*}$ to $\mathcal{E}_{2^*}$.

Consider the problem

$$u = -\lambda\Delta^{-1}(h(\mathbf{x}) - g(\mathbf{x},u)), \qquad u \in \mathcal{E}_{2^*}. \tag{26.9}$$

It is easy to see that if u is a solution to (26.9), then for every $0 \le \lambda \le 1$

$$\|u\|^2_{\mathcal{E}_{2^*}} \le -\lambda \int\limits_{\mathbb{R}^N} u\,g(\mathbf{x},u)\,d\mathbf{x} + \lambda\|h\|_{\mathcal{E}^*_{2^*}}\,\|u\|_{\mathcal{E}_{2^*}} \le \lambda\,\|h\|_{\mathcal{E}^*_{2^*}}\,\|u\|_{\mathcal{E}_{2^*}}.$$

Hence,

$$\|u\|_{\mathcal{E}_{2^*}} \le \mathbf{C}$$

with $\mathbf{C}$ independent of $0 \le \lambda \le 1$, and we can use the Leray-Schauder theorem to complete the proof. □

26.4 BASIC SPACES

In order to obtain more refined existence theorems various weighted spaces and embedding theorems are required. Here we shall consider some Kudrjavcev spaces (Kudrjavcev [1,2], Triebel [1]).

For every $1 < p < +\infty$ and $\delta \in \mathbb{R}$, the closure of $\mathcal{D}(\mathbb{R}^N)$ in the norm

$$\|u\|_{p,(\delta)} = \|(1+|\mathbf{x}|^2)^{\delta/2}u\|_p$$

we denote by $L_{p,(\delta)}$. The closure of $\mathcal{D}(\mathbb{R}^N)$ in the norm

$$\|u\|_{p,(1,\delta)} = \|u\|_{p,(\delta)} + \sum_{i=1}^{N} \|(1+|\mathbf{x}|^2)^{\frac{\delta+1}{2}}\nabla_i u\|_p$$

we denote by $W^1_{p(\delta)}$ and the closure of $\mathcal{D}(\mathbb{R}^N)$ in the norm

$$\|u\|_{p,(2,\delta)} = \|u\|_{p,(1,\delta)} + \sum_{\substack{i=1\\ j=1}}^{N} \|(1+|\mathbf{x}|^2)^{\frac{\delta+2}{2}}\nabla_i\nabla_j u\|_p$$

we denote by $W^2_{p,(\delta)}$.

All these spaces are reflexive.

THEOREM 26.3. *The embedding $W^2_{p,(\delta_1)} \circlearrowleft W^1_{p,(\delta_2)}$ is compact provided $\delta_1 > \delta_2$.*

The following theorems are due to Nirenberg and Walker [1], McOwen [1], and Cantor [1].

THEOREM 26.4. *Let $-N/p < \delta < N-2-N/p$. Then the operator $-\Delta : W^2_{p,(\delta)} \to L_{p,(\delta+2)}$ is an isomorphism, i.e., the equation*

$$\Delta u = f$$

has a unique solution $u \in W^2_{p,(\delta)}$ for any function $f \in L_{p,(\delta+2)}$ and there exist two constants C_1 and C_2 such that for all such u and f

$$C_1 \|u\|_{p,(2,\delta)} \le \|f\|_{p,(\delta+2)} \le C_2 \|u\|_{p,(2,\delta)}.$$

Remark. In contrast to the definition of a generalized solution in Section 2.2, we use here other classes of spaces. Consequently, it is necessary to define separately in what sense the boundary condition is understood. The following result on this account is known.

THEOREM 26.5. *Let $u \in W^1_{p,(\delta)}$ with $N < p$. Then, for any $\gamma > -N/p - \delta$ the estimate $u = O(|\mathbf{x}|^\gamma)$ holds as $|\mathbf{x}| \to \infty$.*

Proof of Theorems 26.4 and 26.5. See McOwen [1] for the proof of Theorem 26.4 and Cantor [1] for the proof of Theorem 26.5. □

The reader may formulate the corresponding existence theorem.

§27. The Method of A Priori Estimates for Radial Solutions

In the following sections, we shall consider the problem

$$\begin{aligned} &-\Delta u + f(r,u) + h(r) = 0, \qquad r \ge 0, \\ &u \in C^2(\mathbb{R}^N), \end{aligned} \tag{27.1}$$

where $f : \mathbb{R}^+ \times \mathbb{R} \to \mathbb{R}$ is a Carathéodory function locally Hölder in the first argument and locally Lipschitz in the second, and $h : \mathbb{R}^+ \to \mathbb{R}$ is a continuous function.

In Chapter 4, it was shown that problem (27.1) can be reduced in the class of radial functions to the Cauchy problem

$$\begin{aligned} &u'' + \frac{N-1}{r}u' - f(r,u) = h(r) \qquad r \ge 0, \\ &u(0) = a, \\ &u'(0) = 0. \end{aligned} \tag{27.2}$$

This problem has a local solution for every a. We shall be interested in possible extensions of solutions and their properties.

27.1 A Theorem on A Priori Estimates

THEOREM 27.1. *Let there exist functions φ_1, φ_2 continuous on $(0,+\infty)$ such that for any function $u \in C^2_{\mathrm{loc}}(\mathbb{R}^N)$ satisfying*

$$u'' + \frac{N-1}{r}u' - f(r,u) = h(r) \quad r \geq 0,$$
$$u'(0) = 0,$$

the estimate

$$\varphi_1(r) \leq u(r) \leq \varphi_2(r), \quad r > 0 \tag{27.3}$$

holds.

Then for any a, problem (27.2) has an entire solution u_a with $u(0) = a$ satisfying (27.3).

Proof. For any $a \in \mathbb{R}$ this problem has a local classical solution $u \in C^2_{\mathrm{loc}}([0,\varepsilon))$ by Theorem 19.1. There is $r_0 < \varepsilon$ such that this solution satisfies a priori estimates (27.3) for $r \geq r_0$. Hence, by Theorem 19.6 the solution can be extended to $[0,+\infty)$, and (27.3) holds for all $r > 0$. □

Remark. Note that *a priori estimates* of solutions to equations are those for which we assume that such solutions exist a priori, although we do not really know if they exist indeed. We have already used some in proofs of previous sections.

In the next section, we demonstrate how, making use of integral relations, one can get such estimates for any $r > 0$ and apply them to the study of problem (27.2) in regard to solvability in different classes defined by these estimates.

27.2 The First Sample

Consider the radial classical solutions, $u \in C^2_{\mathrm{loc}}(\mathbb{R}^N)$, to the Emden-Fowler type equation

$$-A_0(u) \equiv \Delta u + |u|^{p-2}u = h(r) \qquad \text{on } \mathbb{R}^N, \qquad N \geq 3, \tag{27.4}$$

with the exponent $p > 2^*$.

The function $h(r)$ is assumed to be continuous on $\mathbb{R}^+$.

Consider the multiplier

$$M_\alpha u = r\frac{du}{dr} + \alpha\, u$$

with $\alpha \in \mathbb{R}$. Multiply the operator $-A_0$ by M_α and integrate from 0 to r with the weight r^{N-1}. Then we get

$$\begin{aligned}
-\int_0^r & A_0(u(s))\, M_\alpha(u(s)) s^{N-1}\, ds \\
&= r^N\left(\frac{1}{2}|u'(r)|^2 + \frac{1}{p}|u(r)|^p\right) \\
&+ \left(\frac{N}{2} - \alpha - 1\right)\int_0^r |u'(r)|^2 s^{N-1}\, ds \\
&+ \left(\alpha - \frac{N}{p}\right)\int_0^r |u(r)|^p s^{N-1}\, ds + \alpha r^{N-1} u(r)\, u'(r).
\end{aligned} \tag{27.5}$$

27.3 A Priori Estimates

Consider a possible classical radial solution u to equation (27.4) for $h \in C_{\text{loc}}(\mathbb{R}^+)$. For this solution we have by the Young inequality,

$$\left|\int_0^r h(s)\, u'(s)\, s^N\, ds\right| \le \frac{\gamma_1}{2}\int_0^r |u'(s)|^2 s^{N-1}\, ds + \frac{1}{2\gamma_1}\int_0^r h^2(s) s^{N+1}\, ds,$$

$$\left|\alpha\int_0^r h(s)\, u(s)\, s^{N-1}\, ds\right| \le \frac{\gamma_2}{2}\int_0^r |u(s)|^p s^{N-1}\, ds + \gamma_3\int_0^r |h(s)|^{p'} s^{N-1}\, ds$$

with $\gamma_1 > 0, \quad \gamma_2 > 0, \quad \gamma_3 = \dfrac{1}{p'}\left(\dfrac{2}{p\,\gamma_2}\right)^{\frac{1}{p-1}} |\alpha|^{p'}, \quad \dfrac{1}{p} + \dfrac{1}{p'} = 1,$ and

$$|\alpha u(r)\, u'(r) r^{N-1}| \le \frac{1}{4}|u'(r)|^2 r^N + \alpha^2 u^2 r^{N-2}.$$

We make use of these inequalities for special values of α, γ_1, and γ_2.

If $p > 2^*$, $N > 2$, choose $\alpha = \alpha_0$ such that $\frac{N-2}{2} > \alpha_0 > \frac{N}{p}$ and set

$$\bar{\gamma}_1 = \frac{N-2}{2} - \alpha_0 > 0, \qquad \bar{\gamma}_2 = \alpha_0 - \frac{N}{p} > 0.$$

Then taking these inequalities and integral relation (27.5) with $\alpha = \alpha_0$ into account, we get for all possible radial classical solutions to equation (27.4)

$$\begin{aligned}
\frac{1}{4} r^N |u'(r)|^2 &+ \frac{r^N}{p}|u(r)|^p + \frac{\bar{\gamma}_1}{2}\int_0^r |u'(s)|^2 s^{N-1}\, ds + \frac{\bar{\gamma}_2}{2}\int_0^r |u(s)|^p s^{N-1}\, ds \\
&\le \frac{1}{2\bar{\gamma}_1}\int_0^r h^2(s) s^{N+1}\, ds + \bar{\gamma}_3\int_0^r |h(s)|^{p'} s^{N-1}\, ds + \alpha_0^2 r^{N-1} u^2(r).
\end{aligned}$$

From this inequality, by virtue of the estimate

$$\alpha_0^2 r^{N-2} u^2(r) \le \frac{1}{2p} r^N |u(r)|^p + K_1 r^\sigma$$

with $\sigma = \frac{(N-2)p-2N}{p-2} > 0$ for $p > 2^*$ and $K_1 = \frac{p-2}{p} 4^{\frac{2}{p-2}} \alpha_0^{\frac{2p}{p-2}}$, it follows that

$$\frac{1}{4} r^N |u'(r)|^2 + \frac{1}{2p} r^N |u(r)|^p + \frac{\bar{\gamma}_1}{2} \int_0^r |u'(s)|^2 s^{N-1}\, ds + \frac{\bar{\gamma}_2}{2} \int_0^r |u(s)|^p s^{N-1}\, ds \le K_1 r^\sigma + H(r),$$

where

$$H(r) = \frac{1}{\bar{\gamma}_1} \int_0^r h^2(s) s^{N+1}\, ds + \bar{\gamma}_3 \int_0^r |h(s)|^{p'} s^{N-1}\, ds.$$

Hence we get for a local radial classical solution u with $u'(0) = 0$, the following a priori estimates

$$\begin{cases} |u'(r)|^2 \le 4K_1 r^{-\frac{2p}{p-2}} + 4r^{-N} H(r), \\ |u(r)|^p \le 2pK_1 r^{-\frac{2p}{p-2}} + 2pr^{-N} H(r), \\ \dfrac{1}{r^\sigma} \displaystyle\int_0^r |u'(s)|^2 s^{N-1}\, ds \le \dfrac{2K_1}{\bar{\gamma}_1} + \dfrac{2r^{-\sigma}}{\bar{\gamma}_1} H(r), \\ \dfrac{1}{r^\sigma} \displaystyle\int_0^r |u(s)|^p s^{N-1}\, ds \le \dfrac{2K_1}{\bar{\gamma}_2} + \dfrac{2r^{-\sigma}}{\bar{\gamma}_2} H(r) \end{cases} \tag{27.6}$$

for $r \in (0, R]$.

27.4 ASYMPTOTIC ESTIMATES

It is clear that estimates (27.6) remain true for $r \in (0, +\infty)$ for radial classical solutions to (27.4) with $u'(0) = 0$. In this case, as $r \to \infty$ we obtain asymptotic estimates for such possible radial solutions for $p > 2^*$

$$\begin{cases} |u'(r)|^2 \le 4K_1 r^{-\frac{2p}{p-2}} + 4r^{-N} H(r), \\ |u(r)|^p \le 2pK_1 r^{-\frac{2p}{p-2}} + 2pr^{-N} H(r), \\ \dfrac{1}{r^\sigma} \displaystyle\int_0^r |u'(s)|^2 s^{N-1}\, ds \le \dfrac{2K_1}{\bar{\gamma}_1} + \dfrac{2r^{-\sigma}}{\bar{\gamma}_1} H(r), \\ \dfrac{1}{r^\sigma} \displaystyle\int_0^r |u(s)|^p s^{N-1}\, ds \le \dfrac{2K_1}{\bar{\gamma}_2} + \dfrac{2r^{-\sigma}}{\bar{\gamma}_2} H(r) \end{cases} \tag{27.7}$$

as $r \to \infty$.

27.5 UNIMPROVABILITY OF ASYMPTOTIC ESTIMATES

The inequality $p > 2^*$ implies the relation $N > \frac{2p}{p-2}$ for functions h satisfying the inequality

$$H(r) \leq \text{Const} < \infty \qquad \text{for } r \in (0, \infty), \tag{27.8}$$

therefore, the leading, i.e., slower decreasing, term in estimates (27.7) is the first one on the right. This leads to the question of whether the indicated decay exponent for the functions $|u'(r)|$ and $|u(r)|$ as $r \to \infty$ is sharp in this case.

To see this, consider the example

$$\begin{aligned} &\Delta u + |u|^{p-2}u = h_0(|\mathbf{x}|) \qquad \text{on } \mathbb{R}^N, \quad N > 2, \\ &u \to 0 \qquad \text{as } |\mathbf{x}| \to \infty, \end{aligned}$$

with the function

$$h_0(|\mathbf{x}|) = h_0(r) = B_0(1 + r^2)^{-\frac{2p-3}{p-2}}$$

and the solution

$$u(r) = B_1(1 + r^2)^{-\frac{1}{p-2}},$$

where

$$B_1 = 2^{\frac{1}{p-2}}(p-2)^{-\frac{2}{p-2}}[(N-2)(p-2) - 2]^{\frac{1}{p-2}},$$
$$B_0 = -4(p-1)(p-2)^{-2}B_1.$$

Introduce for this example the quantity

$$p_+(N) = \frac{3(N-2) + \sqrt{N^2 - 4N - 36}}{2(N-4)} \qquad \text{for } N > 4$$

and set

$$p_+(N) = +\infty \qquad \text{for } N = 3 \text{ or } N = 4.$$

Then for an exponent p satisfying

$$2^* < p < p_+(N) \qquad (N > 2),$$

the function h_0 satisfies condition (27.8), that is,

$$H_0(r) = \frac{1}{2\bar{\gamma}_1} \int_0^r h_0^2(s) s^{N+1}\, ds + \bar{\gamma}_3 \int_0^r |h_0(s)|^{p'} s^{N-1}\, ds \leq \text{Const} < \infty$$

for $r \in (0, \infty)$, and for the solution u the following limit relations are true:

$$|u'(r)|^2 = 4(p-2)^{-2} B_1^2 r^{-\frac{2p}{p-2}} \left(1 - 2\frac{p-1}{p-2}\frac{1}{r^2} + \cdots\right),$$

$$|u(r)|^p = B_1^p r^{-\frac{2p}{p-2}} \left(1 - \frac{p}{p-2}\frac{1}{r^2} + \cdots\right),$$

$$\frac{1}{r^\sigma}\int_0^r |u'(s)|^2 s^{N-1}\,ds = 4(p-2)^{-2} B_1^2 \int_0^1 \left(t^2 + \frac{1}{r^2}\right)^{-2\frac{p-1}{p-2}} t^{N+1}\,dt$$

$$\to 4(p-2)^{-2} B_1^2 \sigma^{-1} \qquad \text{as } r \to \infty,$$

$$\frac{1}{r^\sigma}\int_0^r |u(s)|^p s^{N-1}\,ds = B_1^p \int_0^1 \left(t^2 + \frac{1}{r^2}\right)^{-\frac{p}{p-2}} t^{N-1}\,dt \to B_1^p \sigma^{-1} \qquad \text{as } r \to \infty.$$

Therefore, exponents in the leading terms of inequalities (27.7) for radial classical solutions to problem (27.4) with $h = h_0$ under condition (27.8) and indicated p are optimal.

Remark. Under the conditions mentioned in the examples we have

$$\int_0^r |u'(s)|^2 s^{N-1}\,ds \to +\infty \qquad \text{and} \qquad \int_0^r |u(s)|^p s^{N-1}\,ds \to +\infty \qquad \text{as } r \to \infty.$$

This means that the solution $u(r) = u(|\mathbf{x}|)$ does not belong to $L_p(\mathbb{R}^N)$ and $H_0^1(\mathbb{R}^N)$. Hence, it is impossible to establish the existence of such solutions by the variational methods in the ordinary (nonweighted) Sobolev spaces.

27.6 Existence of Radial Solutions

Consider the Cauchy problem

$$\begin{aligned} &u'' + \frac{N-1}{r}u' + |u|^{p-2} = h(r) \qquad r \geq 0,\\ &u(0) = u_0,\\ &u'(0) = 0, \end{aligned} \tag{27.9}$$

where $h \in C_{\text{loc}}(\mathbb{R}^+)$, $p > 2^*$, and $N \geq 3$.

THEOREM 27.2. *Problem (27.9), with $p > 2^*$ and $N > 2$, has a classical radial solution u satisfying estimates (27.7) as $r \to \infty$ for all u_0.*

Proof. It follows from Theorem 27.1. □

Now we consider the elliptic problem on $\mathbb{R}^N$,

$$\begin{aligned} &\Delta u + |u|^{p-2}u = h_0(|\mathbf{x}|) \qquad \text{on } \mathbb{R}^N \quad (N > 2),\\ &u \to 0 \qquad \text{as } |\mathbf{x}| \to \infty, \end{aligned} \tag{27.10}$$

with $p > 2^*$, in the class of radial functions. As for h, we suppose that the following condition holds:

CONDITION (H_∞). The function h belongs to $C_{\mathrm{loc}}(\mathbb{R}^+)$ and

$$\lim_{r\to\infty} r^{-N}\left[\int_0^r |h(s)|^2 s^{N+1}\,ds + \int_0^r |h(s)|^{p'} s^{N-1}\,ds\right] = 0.$$

THEOREM 27.3. *Let h satisfy condition (H_∞). Then problem (27.10) has a continuum (a one-parameter family) of classical radial solutions satisfying estimates (27.7) as $r \to \infty$.*

Proof. This theorem follows immediately from Theorem 27.1 and a priori estimates (27.7). □

27.7 SIGN-DEFINITENESS OF SOLUTIONS

In this section, we establish a theorem on the sign of radial solutions and prove on this basis a theorem on the existence of a continuum of positive radial solutions to elliptic problem (27.10).

Assuming that $h \in C^1([0,R])$, we apply integral relation (27.5) for a classical radial solution u of (27.4) with $u'(0) = 0$ and get

$$\begin{aligned} -\alpha r^{N-1} u(r)\, u'(r) &= r^N \left(\frac{|u'(r)|^2}{2} + \frac{|u(r)|^p}{p}\right) \\ &+ \left(\frac{N}{2} - \alpha - 1\right) \int_0^r |u'(s)|^2 s^{N-1}\,ds + \left(\alpha - \frac{N}{p}\right) \int_0^r |u(s)|^p s^{N-1}\,ds \\ &- r^N u(r)\, h(r) + \int_0^r u(s)[sh'(s) + (N-\alpha)h(s)] s^{N-1}\,ds \end{aligned} \tag{27.11}$$

for $0 < r < R$.

THEOREM 27.4. *Let the function $h \in C^1([0,R])$ be such that there exists a constant α such that*

$$\frac{N-2}{2} \ge \alpha \ge \frac{N}{p} \qquad (p > 2^*) \tag{27.12}$$

and

$$-M_\alpha^* h(r) \equiv r h'(r) + (N-\alpha)\, h(r) \ge 0 \tag{27.13}$$

for $0 < r < R$.

Then for any radial classical solution to equation (27.4) with $p > 2^*$ and $u(0) > 0$, the a priori inequality

$$u(r) > 0 \qquad \text{for } 0 < r < R \tag{27.14}$$

follows, and for the derivative of this solution the inequality

$$\alpha u'(r) \le r\,h(r) \qquad \text{for } 0 < r < R \tag{27.15}$$

holds.

Proof. The proof follows immediately from relation (27.11). Indeed, otherwise there exists an interval $[0, R_0] \subset [0, R]$ such that

$$u(r) > 0 \qquad \text{as } 0 < r < R_0 \quad \text{and } u(R_0) = 0.$$

This contradicts relation (27.14) considered on this interval.

Now assertion (27.15) follows from (27.14) by inequality (27.11). □

As a corollary of this theorem we have the following result.

COROLLARY. *Let $h \in C^1_{\text{loc}}(\mathbb{R}^+)$ and*

$$r\,h'(r) + (N - \alpha)\,h(r) \ge 0 \qquad \text{for } 0 < r < \infty, \tag{27.16}$$

where α is given by inequality (27.12). Then for any radial classical solution to equation (27.4) with $p > 2^*$, $N > 2$, and $u(0) > 0$, the a priori inequality

$$u(r) > 0 \qquad \text{for } 0 < r < \infty$$

follows, and for the derivative of this solution the inequality (27.15) holds for $0 < r < \infty$.

THEOREM 27.5. *Let $h \in C^1([0, R])$ and*

(i) the function h satisfy condition (H_∞),

(ii) h satisfy inequality (27.16), where α is from inequality (27.12).

Then problem (27.10), with $p > 2^*$ and $N > 2$, has a continuum of positive radial classical solutions.

27.8 AN EXAMPLE

Consider the equation

$$\Delta u + |u|^{p-2}u = A(1 + \lambda^2|\mathbf{x}|^2)^\gamma \qquad \text{in } \mathbb{R}^N, \quad N \ge 3 \tag{27.17}$$

with $p > 2^*$, $A > 0$, and $\lambda \in \mathbb{R}$. Positivity condition (27.16) is true for all $\gamma \ge -\frac{N}{2p'}$, $\frac{1}{p} + \frac{1}{p'} = 1$.

Indeed, for $\alpha = \frac{N}{p}$ we have

$$\begin{aligned} rh'(r) + (N - \alpha)h(r) &= A[(N - \alpha) + (N - \alpha + 2\gamma)\lambda^2 r^2](1 + \lambda^2 r^2)^{\gamma - 1} \\ &= A\left[\frac{N}{p'} + \left(\frac{N}{p'} + 2\gamma\right)\lambda^2 r^2\right](1 + \lambda^2 r^2)^{\gamma-1} > 0. \end{aligned}$$

Consequently, for any $\gamma \geq -\frac{N}{2p'}$, equation (27.17), with $p > 2^*$, $A > 0$, and $\lambda \in \mathbb{R}$, has a continuum of classical positive radial solutions.

Asymptotic estimates as $r \to \infty$ for these solutions are obtained on the basis of inequalities (27.7).

In particular, Condition (H_∞) for this example is satisfied for $\gamma < -1/2$.

Thus, for any $\gamma : -\frac{1}{2} > \gamma > -\frac{N}{2p'}$, equation (27.17) has a one-parameter continuum of positive radial decaying solutions.

COROLLARY. *The problem*

$$\Delta u + |u|^{p-2}u = 0 \qquad \text{on } \mathbb{R}^N \quad (N > 2),$$
$$u \to 0 \qquad \text{as } |\mathbf{x}| \to \infty,$$

for $p > 2^*$ has a one-parameter continuum of positive radial classical solutions.

Clearly, for this homogeneous problem the existence of a continuum of positive radial solutions follows from the existence of a single positive (nontrivial) radial solution.

27.9 THE SECOND SAMPLE

In this section the results of the previous section are generalized. We consider here the equation

$$\Delta u + q(|\mathbf{x}|)\,|u|^{p-2}u = h(r) \qquad \text{on } \mathbb{R}^N, \qquad N \geq 3 \tag{27.18}$$

with the exponent $p > 2^*$.

27.10 THE VARIATIONAL IDENTITY

Analogously to the previous section, we multiply equation (27.18) by

$$M_\alpha u = r\frac{du}{dr} + \alpha\, u.$$

and integrate from 0 to r with the weight r^{N-1}. Then we get

$$r^N\left[\frac{1}{2}|u'(r)|^2 + \frac{1}{p}q(r)\,|u(r)|^p\right] + \left(\frac{N}{2} - \alpha - 1\right)\int_0^r |u'(r)|^2 s^{N-1}\,ds$$
$$+ \int_0^r \left[\left(\alpha - \frac{N}{p}\right)q(s) - \frac{1}{p}s\,q'(s)\right]|u(r)|^p s^{N-1}\,ds$$
$$= \int_0^r h(s)[su'(s) + \alpha u(s)]\,s^{N-1}ds - \alpha r^{N-1}u(r)\,u'(r). \tag{27.19}$$

27.11 A PRIORI ESTIMATES

We suppose that the function q satisfies the following condition:

CONDITION (q).

(i) $q \in C^1_{\text{loc}}(\mathbb{R}^+)$,

(ii) $q(r) > 0$ for all $r > 0$,

(iii) there exists a number $\alpha_0 < (N-2)/2$ such that

$$q_1(r) = \left(\alpha_0 - \frac{N}{p}\right) q(r) - \frac{1}{p} r q'(r) > 0 \qquad \text{for } r > 0.$$

Remark. It is obvious that this condition is substantial if $N/p \le \alpha_0 < (N-2)/2$.

Denote $\gamma_0 = (N-2)/2 - \alpha_0 > 0$. Then we have for $h \in C_{\text{loc}}(\mathbb{R}^+)$,

$$\left|\int_0^r h(s)\, u'(s)\, s^N\, ds\right| \le \frac{\gamma_0}{2} \int_0^r |u'(s)|^2 s^{N-1}\, ds + \frac{1}{2\gamma_0} H_0(r),$$

$$H_0(r) = \int_0^r s^2 h^2(s)\, s^{N-1} ds;$$

$$\alpha_0 \left|\int_0^r h(s)\, u(s)\, s^{N-1}\, ds\right| \le \frac{1}{2} \int_0^r q_1(s)\, |u(s)|^p s^{N-1}\, ds + \gamma_1 H_1(r),$$

$$\gamma_1 = \frac{|\alpha_0|^{p'}}{p'} \left(\frac{2}{p}\right)^{\frac{1}{p-1}}, \qquad \frac{1}{p} + \frac{1}{p'} = 1,$$

$$H_1(r) = \int_0^r [q_1(s)]^{-1/(p-1)} |h(s)|^{p'} s^{N-1}\, ds.$$

We have also

$$|\alpha_0 u(r)\, u'(r)\, r^{N-1}| \le r^N \frac{1}{4} |u'(r)|^2 + \frac{r^N}{2p} q(r)\, |u(r)|^p + \gamma_2 q_2(r),$$

$$\gamma_2 = \frac{p-2}{p} 2^{4/(p-2)} |\alpha_0|^{2p/(p-2)}, \qquad p > 2^*,$$

$$q_2(r) = r^{\frac{(N-2)p-2N}{p-2}} [q(r)]^{-2/(p-2)}.$$

Taking into account these inequalities we get from integral relation (27.19) with $\alpha = \alpha_0$,

$$\frac{1}{2} r^N \left[\frac{1}{2}|u'(r)|^2 + \frac{1}{p} q(r)\, |u(r)|^p\right] + \frac{\gamma_0}{2} \int_0^r |u'(s)|^2 s^{N-1}\, ds$$
$$+ \frac{1}{2} \int_0^r q_1(s)\, |u(s)|^p s^{N-1}\, ds \le \gamma_2 q_2(r) + \frac{1}{2\gamma_0} H_0(r) + \gamma_1 H_1(r). \quad (27.20)$$

Thus, the following result is established:

THEOREM 27.6. *Let $N > 2$ and $p > 2^*$. Let a function q satisfy Condition (q). Then for all classical radial solutions u of equation (27.19) estimate (27.20) holds.*

27.12 SHARPNESS OF THE ESTIMATE

The estimate of the decreasing rate for classical radial solutions regarding both the coefficient q and the right-hand side h follows from inequality (27.20). It turns out that this estimate can not be improved under the general conditions of Theorem 27.6.

To show this, consider the example

$$\begin{aligned} &\Delta u + q(r)\,|u|^{p-2}u = 0 \qquad \text{on } \mathbb{R}^N \quad (N > 2),\\ &p > 2\frac{N+2}{N-2} > 2^*,\\ &q(r) = \frac{4\lambda^2 N}{p-2}\left[1 + \frac{(N-2)p-2N}{(p-2)N}\lambda^2 r^2\right], \qquad \lambda \neq 0. \end{aligned} \tag{27.21}$$

For this equation all the assumptions of Theorem 27.6 hold, hence for any classical radial solution of this equation inequality (27.20) is valid with an appropriate function q_2,

$$q_2(r) = \text{Const}\ r^{\frac{(N-2)p-2(N+2)}{p-2}}(1+o(1)) \qquad \text{as } r \to \infty$$

and $H_0(r) \equiv H_1(r) \equiv 0$.

From (27.20) we find that

$$\begin{aligned} &|u(r)| \le C_1 r^{-4/(p-2)}(1+o(1)),\\ &|u'(r)| \le C_2 r^{-(p+2)/(p-2)}(1+o(1)),\\ &\int_0^r |u'(s)|^2 s^{N-1}ds \le C_3 r^{\frac{(N-2)p-2(N+2)}{p-2}}(1+o(1)),\\ &\int_0^r q_1(s)\,|u(s)|^p s^{N-1}ds \le C_4 r^{\frac{(N-2)p-2(N+2)}{p-2}}(1+o(1)), \end{aligned} \tag{27.22}$$

with positive constants C_1, C_2, C_3, and C_4 as $r \to \infty$, where $q_1(r) = a_0 + b_0 r^2$ with appropriate coefficients $a_0 > 0$ and $b_0 > 0$.

On the other hand, the function

$$u(r) = (1 + \lambda^2 r^2)^{-2/(p-2)}$$

is a solution to (27.21) for which estimates (27.22) are unimprovable.

27.13 DECAYING SOLUTIONS

Now we consider conditions under which a solution to (27.18) is decaying, i.e.,

$$\lim_{r\to\infty} u(r) = 0.$$

CONDITION (q_∞). The function q belongs to $C_{\text{loc}}(\mathbb{R}^+)$ and

$$\lim_{r\to\infty} r^2 q(r) = +\infty.$$

CONDITION (H_∞). The function h belongs to $C_{\text{loc}}(\mathbb{R}^+)$ and

$$\lim_{r\to\infty} \frac{H_0(r) + H_1(r)}{r^N q(r)} = 0,$$

where H_0 and H_1 were defined earlier and q satisfies Conditions (q) and (q_∞).

Directly from inequality (27.20) we deduce the following result.

THEOREM 27.7. *Let $N > 2$ and $p > 2^*$. Let the function q satisfy Conditions (q) and (q_∞). Let the function h satisfy Condition (H_∞) with this function q.*

Then any classical radial solution u to equation (27.18) is decaying.

27.14 EXISTENCE OF ENTIRE SOLUTIONS

THEOREM 27.8. *Suppose $N \geq 3$ and $p > 2^*$. Suppose that the function q satisfies Condition (q) and the function h belongs to the class $C_{\text{loc}}(\mathbb{R}^+)$. Then there is a continuum of entire classical radial solutions of equation (27.18). Moreover, if additionally q satisfies Condition (q_∞) and h satisfies Condition (H_∞) with the same function q, then any such solution is decaying.*

Proof. It follows from Theorems 27.1 and 27.7. □

27.15 POSITIVE SOLUTIONS

Let now $q \in C^1_{\text{loc}}(\mathbb{R}^+)$ and $h \in C^1_{\text{loc}}(\mathbb{R}^+)$. Then for any classical radial solution to (27.18) the following holds by (27.19):

$$r^N\left[\frac{|u'(r)|^2}{2} + \frac{q(r)\,|u(r)|^p}{p}\right] + \left(\frac{N}{2} - \alpha - 1\right)\int_0^r |u'(s)|^2 s^{N-1}\,ds$$
$$+ \int_0^r \left[\left(\alpha - \frac{N}{p}\right) q(s) - \frac{1}{p} s q'(s)\right] |u(s)|^p s^{N-1}\,ds + \alpha\, r^{N-1} u(r)\, u'(r)$$
$$= r^N u(r)\, h(r) + \int_0^r u(s)[(\alpha - N)h(s) - s h'(s)] s^{N-1}\,ds$$

for $r > 0$. This identity implies

THEOREM 27.9. *Let $N > 2$ and $p > 2^*$. Suppose that the functions q and h belong to the class $C^1_{\mathrm{loc}}(\mathbb{R}^+)$. Let $q(r) > 0$ for all $r > 0$ and let there exist $\alpha < (N-2)/2$ such that*

$$\left(\alpha - \frac{N}{p}\right) q(r) - \frac{1}{p} r\, q'(r) \geq 0 \qquad \text{for } r > 0,$$
$$r h'(r) + (N - \alpha)\, h(r) \geq 0 \qquad \text{for } r > 0. \tag{27.23}$$

Then for any classical radial solution to equation (27.18) with $u(0) > 0$ the inequality

$$u(r) > 0 \qquad \text{for } r > 0$$

follows, and for the derivative of this solution the inequality

$$\alpha u'(r) \leq r\, h(r) \qquad \text{for } r > 0$$

holds.

Proof. It is analogous to the proof of Theorem 27.4. □

THEOREM 27.10. *Let $N > 2$ and $p > 2^*$. Let q satisfy Condition (q). Let h belong to $C^1_{\mathrm{loc}}(\mathbb{R}^+)$ and satisfy inequality (27.23) with $\alpha = \alpha_0$ from Condition (q).*

Then there exists a continuum of classical positive radial solutions to equation (27.18).

Moreover, if additionally the function q satisfies Condition (q_∞) and the function h satisfies Condition (H_∞) with this function q, then any such solution is decaying.

27.16 EXAMPLE

Consider the equation

$$\Delta u + (1 + \lambda^2 r^2)^\nu |u|^{p-2} u = A(1 + \mu^2 |\mathbf{x}|^2)^\sigma, \qquad \text{in } \mathbb{R}^N, \qquad N > 2, \tag{27.24}$$

with $p > 2^*$, $A > 0$, where $\lambda \neq 0$, $\mu \neq 0$, σ, ν are real parameters. Then if

$$\max\left(\frac{N}{p}, \frac{N + 2\nu}{p}\right) < \min\left(\frac{N-2}{2}, N + 2\sigma\right), \tag{27.25}$$

equation (27.24) has a continuum of classical positive radial solutions.

Moreover, if

$$\nu > \max(-1, \sigma, 1 + 2\sigma), \tag{27.26}$$

the additional assumptions of Theorem 27.10 are true. Then, under conditions (27.25) and (27.26), equation (27.24) has a continuum of classical positive decaying radial solutions.

27.17 THE THIRD SAMPLE

In this section we shall dwell upon a general case

$$\begin{aligned} -A(u) &\equiv \Delta u + f(r,u) = h(r) \qquad \text{in } \mathbb{R}^N, \qquad N > 2, \\ u &\in C^2_{\mathrm{loc}}(\mathbb{R}^N). \end{aligned} \tag{27.27}$$

We assume that functions f and F_r belong to $C_{\mathrm{loc}}(\mathbb{R}^+ \times \mathbb{R})$ and $h \in C_{\mathrm{loc}}(\mathbb{R}^+)$.

Observe that the form of (27.27) with a separate function h is dictated by the necessity of splitting off the properly nonlinear operator and the corresponding right-hand side from the range of this operator.

Consider the integral multiplier

$$M_\alpha u = r\frac{du}{dr} + \alpha\, u,$$

again. Multiply the operator $-A$ by M_α and integrate from 0 to r with the weight r^{N-1}.

Then we get

$$\begin{aligned} -\int_0^r A(u(s))\, M_\alpha(u(s)) s^{N-1}\, ds &= r^N \left(\frac{1}{2}|u'(r)|^2 + F(r,u(r))\right) + \alpha r^{N-1} u(r)\, u'(r) \\ &+ \left(\frac{N}{2} - \alpha - 1\right) \int_0^r |u'(r)|^2 s^{N-1}\, ds \\ &+ \int_0^r [\alpha\, u\, f(s,u) - N\, F(s,u) - s\, F_s(s,u)] s^{N-1}\, ds. \end{aligned} \tag{27.28}$$

Here

$$F(s,t) = \int_0^t f(s,t')\, dt' \qquad \text{and} \qquad F_s(s,t) = \frac{\partial F(s,t)}{\partial s}.$$

27.18 A PRIORI ESTIMATES

Consider a possible classical radial solution to (27.27). For this solution the inequalities of Section 27.3 are valid with $p > 2$:

$$\left|\int_0^r h(s)\, u'(s)\, s^N\, ds\right| \le \frac{\gamma_1}{2} \int_0^r |u'(s)|^2 s^{N-1}\, ds + \frac{1}{2\gamma_1} \int_0^r h^2(s) s^{N+1}\, ds,$$

$$\alpha \left|\int_0^r h(s)\, u(s)\, s^{N-1}\, ds\right| \le \frac{\gamma_2}{2} \int_0^r |u(s)|^p s^{N-1}\, ds + \gamma_3 \int_0^r |h(s)|^{p'} s^{N-1}\, ds$$

with $\gamma_1 > 0$, $\gamma_2 > 0$, $\gamma_3 = \frac{1}{p'\,\gamma_2}\left(\frac{2}{p\gamma_2}\right)^{\frac{1}{p-1}}|\alpha|^{p'}$, $\frac{1}{p}+\frac{1}{p'} = 1$, and $|\alpha u(r)\,u'(r)r^{N-1}| \leq \frac{1}{4}|u'(r)|^2r^N + \alpha^2u^2(r)r^{N-2}$.

Then we get from integral relation (27.28)

$$r^N\left(\frac{1}{2}|u'(r)|^2 + F(r,u(r))\right) + \left(\frac{N}{2}-\alpha-1\right)\int_0^r |u'(r)|^2s^{N-1}\,ds + \int_0^r [\alpha\, u\, f(s,u) - N\,F(s,u) - s\,F_s(s,u)]s^{N-1}\,ds$$
$$\leq \frac{1}{4}r^N|u'(r)|^2 + \alpha^2r^{N-2}|u(r)|^2 + \frac{\gamma_1}{2}\int_0^r |u'(s)|^2s^{N-1}\,ds + \frac{\gamma_2}{2}\int_0^r |u(s)|^ps^{N-1}\,ds + \frac{1}{2\gamma_1}\int_0^r h^2(s)s^{N+1}\,ds + \gamma_3\int_0^r |h(s)|^{p'}s^{N-1}\,ds.$$

From this and the estimate

$$\alpha^2r^{N-2}u^2(r) \leq \frac{1}{2p}r^N|u(r)|^p + K_1r^\sigma \tag{27.29}$$

with $\sigma = \dfrac{(N-2)p-2N}{p-2}$ and $K_1 = \dfrac{p-2}{p}4^{\frac{2}{p-2}}\alpha^{\frac{2p}{p-2}}$, we obtain

$$\frac{1}{4}r^N|u'(r)|^2 + r^N\left[F(r,u(r)) - \frac{1}{2p}|u(r)|^p\right] + \left(\frac{N}{2}-\alpha-1-\frac{\gamma_1}{2}\right)\int_0^r |u'(r)|^2s^{N-1}\,ds + \int_0^r \left[\alpha\, u\, f(s,u) - \frac{\gamma_2}{2}|u|^p - N\,F(s,u) - s\,F_s(s,u)\right]s^{N-1}\,ds$$
$$\leq \frac{1}{2\gamma_1}\int_0^r h^2(s)s^{N+1}\,ds + \gamma_3\int_0^r |h(s)|^{p'}s^{N-1}\,ds + K_1r^\sigma \tag{27.30}$$

for $0 < r < \infty$.

Assume now that the function f satisfies the following condition:

CONDITION (f). The functions f and F_r belong to $C_{\rm loc}(\mathbb{R}^+ \times \mathbb{R})$ and there exist $p > 2$, $\gamma_1 > 0$, $\gamma_2 > 0$, and real α such that

$$\begin{aligned} F(s,t) - \frac{1}{2p}|t|^p &\geq 0, \\ \frac{N}{2} - \alpha - 1 - \frac{\gamma_1}{2} &\geq 0, \\ \alpha\, t\, f(s,t) - \frac{\gamma_2}{2}|t|^p - N\,F(s,t) - s\,F_s(s,t) &\geq 0 \end{aligned} \tag{27.31}$$

for all $s \in \mathbb{R}^+$ and $t \in \mathbb{R}$.

EXAMPLE. Let $f(t) = |t|^{p-2}t$. Then inequalities (27.31) assume the form

$$F(s,t) - \frac{1}{2p}|t|^p \equiv \frac{1}{2p}|t|^p \geq 0, \quad \frac{N}{2} - \alpha - 1 - \frac{\gamma_1}{2} \geq 0,$$

$$\alpha\, t\, f(s,t) - \frac{\gamma_2}{2}|t|^p - N\,F(s,t) - s\,F_s(s,t) \equiv \left(\alpha - \frac{\gamma_2}{2} - \frac{N}{p}\right)|t|^p \geq 0.$$

Thus, in this case, Condition (f) is equivalent to the following: the exponent $p > 2$ is such that there exist $\gamma_1 > 0$, $\gamma_2 > 0$, and real α for which

$$\frac{N}{2} - 1 \geq \alpha + \frac{\gamma_1}{2} \qquad \text{and } \alpha \geq \frac{N}{p} + \frac{\gamma_2}{2},$$

which is equivalent to

$$\frac{N}{2} - 1 > \alpha > \frac{N}{p},$$

that is, $N > 2$ and $p > 2^*$.

Immediately from inequality (27.30) we get the following statement on a priori estimates for radial solutions to equation (27.27).

THEOREM 27.11. *Let f satisfy Condition (f) and h belong to $C_{\rm loc}(\mathbb{R}^+)$. Then for all classical radial solutions u to (27.27) the inequality*

$$r^N|u'(r)|^2 \leq \frac{2}{\gamma_1}\int_0^r h^2(s)s^{N+1}\,ds + \gamma_3\int_0^r |h(s)|^{p'}s^{N-1}\,ds + K_1 r^\sigma \tag{27.32}$$

holds for $0 < r < \infty$.

Here $\gamma_3 = \frac{1}{p'}\left(\frac{2}{p\,\gamma_2}\right)^{\frac{1}{p-1}}|\alpha|^{p'}$, and K_1, σ are defined in (27.29).

Remark. It is clear that by modifying Condition (f), on the basis of (27.28), sharper a priori asymptotic estimates can be obtained for radial classical solutions to equation (27.27) for appropriate functions $f(r,u)$.

27.19 ENTIRE SOLUTIONS

Consider the problem

$$\begin{aligned} &u'' + \frac{N-1}{r}u' + f(r,u) = h(r) \qquad r \geq 0, \\ &u(0) = u_0, \\ &u'(0) = 0 \end{aligned} \tag{27.33}$$

for $N \geq 3$.

Impose the following condition:

CONDITION (H_*). The function h belongs to $C_{\mathrm{loc}}(\mathbb{R}^+)$ and there exist constants $C_1 > 0$, $C_2 > 0$, and $\varepsilon > 0$ such that the inequality

$$\int_0^r h^2(s)s^{N+1}\,ds + \int_0^r |h(s)|^{p'} s^{N-1}\,ds \leq C_1 r^{N-2-\varepsilon} + C_2$$

holds for all $0 < r < \infty$.

THEOREM 27.12. *Let the function $f(r,t)$ satisfy Condition (f) and be locally Lipschitz with respect to $t \in \mathbb{R}$ on $\mathbb{R}^+ \times \mathbb{R}$. Let the function h satisfy Condition (H_*). Then for any $u_0 \in \mathbb{R}$ problem (27.33) has a classical solution, for which the following finite limit exists:*

$$\lim_{r\to\infty} u(r) = u_\infty \neq \infty.$$

Proof. Existence of a solution satisfying (27.32) follows from Theorem 27.1. By condition (H_*), this inequality implies convergence of the integral $\int_{a>0}^{\infty} |u'(r)|\,dr$ and, consequently, the existence of a finite u_∞. □

Now consider additional conditions on f under which $u_\infty = 0$.

27.20 DECAYING SOLUTIONS

Consider the problem

$$\begin{aligned} &\Delta u + f(r,u) = h(r) \qquad \text{in } \mathbb{R}^N, \quad N \geq 3, \\ &u \to 0 \qquad \text{as } r \to \infty, \end{aligned} \tag{27.34}$$

in the class of radial functions.

THEOREM 27.13. *Let the function $f(r,t)$ satisfy Condition (f) and be locally Lipschitz with respect to $t \in \mathbb{R}$ on $\mathbb{R}^+ \times \mathbb{R}$, and satisfy the following limit condition:*

$$\text{if} \quad F(s,t) - \frac{1}{2p}|t|^p \to 0 \quad \text{and} \quad s \to \infty, \quad \text{then } t \to \infty\,. \tag{27.35}$$

Let the function h satisfy condition (H_∞).

Then elliptic problem (27.34) has a continuum of radial classical solutions.

Proof. The existence of a solution follows from Theorem 27.12.

Inequality (27.30) implies, from condition (H_∞),

$$F(r,u) - \frac{1}{2p}|u|^p \to 0 \quad \text{as } r \to \infty$$

for this solution u.

Then decay follows from condition (27.35). □

27.21 SIGN-DEFINED SOLUTIONS

In this section we shall establish a theorem on the sign of radial solutions and prove on this basis a theorem on the existence of a continuum of positive radial solutions to elliptic problem (27.34).

Under the assumption $h \in C^1_{\mathrm{loc}}(\mathbb{R}^+)$ rewrite integral identity (27.28) for a classical radial solution u with $u'(0) = 0$ in the form

$$\begin{aligned} -\alpha\, r^{N-1} u(r)\, u'(r) &= r^N \left(\frac{|u'(r)|^2}{2} + F(r, u(r)) \right) \\ &\quad + \left(\frac{N}{2} - \alpha - 1 \right) \int_0^r |u'(s)|^2 s^{N-1}\, ds \\ &+ \int_0^r [\alpha\, u\, f(s,u) - N\, F(s,u) - s\, F_s(s,u)] s^{N-1}\, ds - r^N\, u(r)\, h(r) \\ &\quad + \int_0^r u(s)[s h'(s) + (N-\alpha) h(s)] s^{N-1}\, ds \end{aligned} \tag{27.36}$$

for $r > 0$.

CONDITION (f_+). The functions f and F_r belong to $C_{\mathrm{loc}}(\mathbb{R}^+ \times \mathbb{R})$ and there exists $\alpha < (N-2)/2$ such that

$$\alpha\, t\, f(s,t) - N\, F(s,t) - s\, F_s(s,t) \geq 0$$

for all $s \in \mathbb{R}^+$ and $t \in \mathbb{R}$.

CONDITION (h_+). The function h belongs to $C^1_{\mathrm{loc}}(\mathbb{R}^+)$ and

$$r h'(r) + (N - \alpha)\, h(r) \geq 0$$

for $r > 0$ where α is from condition (f_+).

THEOREM 27.14. *Let the function f satisfy condition (f_+) and the function h satisfy Condition (h_+).*

Then, for any classical solution to equation (27.34) with $u(0) > 0$ the a priori inequality

$$u(r) > 0 \qquad \text{for } r > 0$$

follows.

If, additionally,

$$F(s,t) \geq 0 \qquad \text{for all } s \in \mathbb{R}^+,\ t \in \mathbb{R}^+$$

for the derivative of this solution the inequality

$$\alpha\, u'(r) \leq r\, h(r) \qquad \text{for } r > 0$$

holds.

Proof. It is analogous to the proof of Theorem 27.4. □

An analogous result is valid for negative solutions.

CONDITION (h_-). The function h belongs to $C^1_{\text{loc}}(\mathbb{R}^+)$ and

$$rh'(r) + (N - \alpha)\, h(r) \leq 0$$

for $r > 0$, where α is from Condition (f_+).

THEOREM 27.15. *Let the function f satisfy Condition (f_+) and the function h satisfy Condition (h_-).*

Then, for any classical solution to equation (27.34) with $u(0) > 0$ the a priori inequality

$$u(r) < 0 \qquad \text{for } r > 0$$

follows.

If, additionally,

$$F(s,t) \geq 0 \qquad \text{for all } s \in \mathbb{R}^+,\ t \in \mathbb{R}^-$$

for the derivative of this solution the inequality

$$\alpha\, u'(r) \geq r\, h(r) \qquad \text{for } r > 0$$

holds.

Thus, we have

THEOREM 27.16. *Let the functions f and h satisfy the conditions of Theorems 27.13 and 27.14. Then problem (27.34) has a continuum of positive classical solutions.*

THEOREM 27.17. *Let the functions f and h satisfy the conditions of Theorems 27.13 and 27.15. Then problem (27.34) has a continuum of negative classical solutions.*

27.22 The Fourth Sample

Let us study classical radial solutions to the equation

$$\Delta u + |u|^{p-2}u = h(r) \qquad \text{on } \mathbb{R}^N, \qquad N \geq 3, \tag{27.37}$$

where the condition $p > 2^*$ is eliminated.

The function $h(r)$ is supposed to be continuous on $\mathbb{R}^+$.

27.23 Variational Identities

Consider the integral multiplier

$$M_\alpha u = r\frac{du}{dr} + \alpha\, u.$$

Multiply equation (27.37) by M_α and integrate from 0 to r with the weight r (instead of r^{N-1} as earlier).

Then we get

$$\begin{aligned} r^2 \left(\frac{1}{2}|u'(r)|^2 + \frac{1}{p}|u(r)|^p\right) & \\ + \alpha r\, u(r)\, u'(r) + \frac{\alpha}{2}(N-2)\,|u(r)|^2 & \\ + (N-\alpha-2)\int_0^r |u'(r)|^2 s\, ds + \left(\alpha - \frac{2}{p}\right)\int_0^r |u(r)|^p s\, ds & \\ = \frac{\alpha}{2}(N-2)\,|u(0)|^2 + \int_0^r h(s)[su'(s) + \alpha u(s)]\, s\, ds. & \end{aligned} \tag{27.38}$$

27.24 A Priori Estimates

Consider a possible classical radial solution $u(r)$ to equation (27.37) provided $h \in C_{\text{loc}}(\mathbb{R}^+)$. For $N > 2$ and $p > 2/(N-2)$ choose $\alpha = \alpha_0$ in such a way that

$$N - 2 > \alpha_0 > 2/p,$$

and denote

$$\gamma_1 = N - 2 - \alpha_0 > 0, \qquad \gamma_2 = \alpha_0 - 2/p > 0.$$

Then we have

$$\left|\int_0^r h(s)\,u'(s)\,s^2\,ds\right| \le \frac{\gamma_1}{2}\int_0^r |u'(s)|^2 s\,ds + \frac{1}{2\gamma_1}\int_0^r h^2(s)s^3\,ds,$$

$$\alpha_0\left|\int_0^r h(s)\,u(s)\,s\,ds\right| \le \frac{\gamma_2}{2}\int_0^r |u(s)|^p s\,ds + \gamma_3\int_0^r |h(s)|^{p'} s\,ds,$$

$$\gamma_3 = \frac{1}{p'}\left(\frac{2}{p\,\gamma_2}\right)^{\frac{1}{p-1}}\alpha_0^{p'}, \quad \frac{1}{p}+\frac{1}{p'} = 1, \qquad \text{and}$$

$$|\alpha_0 r\,u(r)\,u'(r)| \le \frac{\varepsilon}{2}|u'(r)|^2 r^2 + \frac{\alpha_0^2 u^2(r)}{2\varepsilon}, \qquad \varepsilon > 0.$$

Making use of these inequalities with $\alpha = \alpha_0$ and integral relation (27.38), we get

$$\frac{1-\varepsilon_0}{2} r^2 |u'(r)|^2 + \frac{1}{p} r^2 |u(r)|^p + \frac{\alpha_0}{2}\left(N-2-\frac{\alpha_0}{\varepsilon_0}\right)|u(r)|^2$$
$$+\frac{\gamma_1}{2}\int_0^r |u'(s)|^2 s\,ds + \frac{\gamma_2}{2}\int_0^r |u(s)|^p s\,ds \le \frac{\alpha_0}{2}(N-2)|u(0)|^2 + H(r) \quad (27.39)$$

for $r > 0$, where

$$1 > \varepsilon_0 > \frac{\alpha_0}{N-2},$$

$$H(r) = \frac{1}{\gamma_1}\int_0^r h^2(s)s^3\,ds + \gamma_3\int_0^r |h(s)|^{p'} s\,ds.$$

Thus, the following result is established.

THEOREM 27.18. *Let $N > 2$ and $p > 2/(N-2)$. Then for all possible classical radial solutions u to equation (27.37) with $h \in C_{\text{loc}}(\mathbb{R}^+)$, inequality (27.39) holds.*

Let function h satisfy the following condition:

CONDITION (H). The function h is continuous on $\mathbb{R}^+$ and

$$\lim_{r\to\infty} r^{-2}\left[\int_0^r |h(s)|^2 s^3\,ds + \int_0^r |h(s)|^{p'} s\,ds\right] = 0.$$

THEOREM 27.19. *Let $N \geq 3$, $p > 2/(N-2)$, and h satisfy condition (H). Then any classical radial solution to (27.34) is decaying.*

Now let h satisfy a more restrictive condition.

CONDITION (H_1). The function h is continuous on $\mathbb{R}^+$ and

$$\lim_{r\to\infty}\left[\int_0^r |h(s)|^2 s^3\,ds + \int_0^r |h(s)|^{p'} s\,ds\right] < +\infty.$$

THEOREM 27.20. *Let $N \geq 3$, $p > 2/(N-2)$, and h satisfy condition (H_1). Then any classical radial solution u to (27.34) satisfies*

$$\begin{aligned} &\overline{\lim_{r\to+\infty}}\, r^2|u'(r)|^2 < +\infty, \\ &\overline{\lim_{r\to+\infty}}\, r^2|u(r)|^p < +\infty, \end{aligned} \tag{27.40}$$

and

$$\begin{aligned} &\int_0^\infty |u'(s)|^2\, s\,ds < \infty, \\ &\int_0^\infty |u(s)|^p\, s\,ds < \infty. \end{aligned} \tag{27.41}$$

Proof. It follows directly from inequality (27.39). □

Thus, we have

THEOREM 27.21. *Let $N > 2$, and $p > 2$ provided $N = 3$, and $p > 1$ provided $N > 3$. Suppose that $h \in C_{\text{loc}}(\mathbb{R}^+)$.*

Then there exists a continuum of entire classical radial solutions to (27.34). Moreover, if h satisfies (H), then every solution u is decaying; if h additionally satisfies (H_1), then every u satisfies (27.40) and (27.41).

Proof. It follows from Theorem 27.1. □

27.25 POSITIVE SOLUTIONS

We consider equation (27.34) with $h \equiv 0$; that is,

$$\Delta u + |u|^{p-2}u = 0 \qquad \text{in } \mathbb{R}^N, \quad N \geq 3. \tag{27.42}$$

We know that for $p > 2^*$ this equation has a continuum of positive decaying radial classical solutions.

From Gidas and Spruck [1] it is known that for $2 \le p < 2^*$ equation (27.42) does not have a nontrivial nonnegative classical solution. Consequently, all entire classical solutions to (27.42) are of variable sign. Hence, in this case equation (27.42) has a continuum of entire classical solutions of variable sign.

We now consider equation (27.34) with $h \not\equiv 0$. For this equation in the class of radial functions consider the Cauchy problem

$$\begin{aligned} &u'' + \frac{N-1}{r}u' + |u|^{p-2}u = h(r) \qquad r \ge 0, \\ &u(0) = 0, \\ &u'(0) = 0 \end{aligned} \tag{27.43}$$

for $N \ge 3$.

Suppose now that $h \in C^1_{\text{loc}}(\mathbb{R}^+)$. Then, by (27.38), the existing classical radial solution u to (27.43) satisfies

$$\begin{aligned} r^2\left(\frac{|u'(r)|^2}{2} + \frac{|u(r)|^p}{p}\right) &+ \alpha\, r\, u(r)\, u'(r) + \frac{\alpha}{2}(N-2)u^2(r) \\ &+ (N-\alpha-2)\int_0^r |u'(s)|^2 s\,ds + \left(\alpha - \frac{2}{p}\right)\int_0^r |u(s)|^p s\,ds \\ &= r^2 u(r)\, h(r) + \int_0^r u(s)[(\alpha-2)h(s) - sh'(s)]s\,ds \end{aligned} \tag{27.44}$$

as $r > 0$.

THEOREM 27.22. *Let $N \ge 3$ and $p > 2$ for $N = 3$ and $p > 1$ for $N \ge 4$, and suppose that $h \in C^1_{\text{loc}}(\mathbb{R}^+)$ with $h(0) > 0$ is such that there exists α, $2/p \le \alpha \le N-2$, for which*

$$(\alpha - 2)\, h(r) - r\, h'(r) \le 0$$

for $r > 0$. Then any classical solution u to equation (27.43) is positive, that is,

$$u(r) > 0 \qquad \text{for } r > 0.$$

Proof. It may be argued with the help of (27.44) analogously to the proof of Theorem 27.4. □

Thus, we have

THEOREM 27.23. *Let the conditions of Theorem 27.22 be true. Then equation (27.34) has a positive entire radial solution u. If, additionally, h satisfies Condition (H), then this u is decaying.*

If, additionally, h satisfies Condition (H_1), then u satisfies limiting conditions (27.40) and inequalities (27.41).

27.26 EXAMPLE

Consider the equation

$$\Delta u + |u|^{p-2}u = A(1 + \lambda^2|\mathbf{x}|^2)^{\nu} \tag{27.45}$$

with $p > 2$ for $N = 3$ and $p > 1$ for $N \geq 4$; here $A > 0$, ν, and $\lambda \in \mathbb{R}$ are constants. If $\nu \geq -1/p'$, there exists a positive classical radial solution to equation (27.45). Moreover, if $p > 2$ for $N = 3$ and $p \geq 2$ for $N \geq 4$, and $-1/p' \leq \nu \leq -1/2$, there exists a positive decaying classical radial solution to (27.45).

27.27 THE FIFTH SAMPLE

Finally, radial classical solutions to the equation

$$\Delta u + f(|\mathbf{x}|, u) = 0 \qquad \text{on } \mathbb{R}^N, \qquad N \geq 3 \tag{27.46}$$

are considered with the help of the most general identities.

We assume that functions f and F_r belong to $C_{\text{loc}}(\mathbb{R}^+ \times \mathbb{R})$.

27.28 THE VARIATIONAL IDENTITY

Recall that aforesaid results concerning this problem were obtained on the basis of some identities.

The first class was

$$\begin{aligned}\int_0^r [\Delta u + f(s,u)] M_\alpha(u(s)) s^{N-1}\, ds &= r^N \left(\frac{1}{2}|u'(r)|^2 + F(r, u(r))\right) \\ &\quad + \alpha r^{N-1} u(r)\, u'(r) \\ &\quad + \left(\frac{N-2}{2} - \alpha\right) \int_0^r |u'(s)|^2 s^{N-1}\, ds \\ &\quad + \int_0^r [\alpha u f(s,u) - N\, F(s,u) - sF_s(s,u)]\, s^{N-1} ds,\end{aligned}$$

where

$$M_\alpha u = r\frac{du}{dr} + \alpha\, u.$$

The second class was

$$\int_0^r [\Delta u + f(s,u)] M_\alpha(u(s)) s\, ds$$
$$= r^2 \left(\frac{1}{2}|u'(r)|^2 + F(r,u(r))\right)$$
$$+ \alpha r\, u(r)\, u'(r) + \frac{\alpha}{2}(N-2)\,|u(r)|^2$$
$$- \frac{\alpha}{2}(N-2)\, u^2(0) + (N-2-\alpha) \int_0^r |u'(s)|^2 s\, ds$$
$$+ \int_0^r [\alpha u f(s,u) - 2F(s,u) - sF_s(s,u)]\, s\, ds. \quad (27.47)$$

These identities were the results of multiplication of the original equation by $M_\alpha u$ and integrating by parts with weights r^{N-1} or r. Now we consider the general weight r^k, $k > 1$, and get the third class of identities.

$$\int_0^r [\Delta u + f(s,u)] M_\alpha(u(s)) s^k\, ds$$
$$= r^{k+1} \left(\frac{1}{2}|u'(r)|^2 + F(r,u(r))\right)$$
$$+ \alpha r^k u(r)\, u'(r) + \frac{\alpha}{2}(N-k-1) r^{k-1} |u(r)|^2$$
$$+ \left(N - 1 - \alpha - \frac{k+1}{2}\right) \int_0^r |u'(s)|^2 s^k\, ds$$
$$+ \int_0^r [\alpha u f(s,u) - (k+1)F(s,u) - sF_s(s,u)]\, s^k ds$$
$$- \alpha \frac{(N-k-1)(k-1)}{2} \int_0^r |u(s)|^2\, s^{k-2} ds. \quad (27.48)$$

Later on we shall use these relations.

27.29 A PRIORI ESTIMATES

Consider the equation

$$\begin{aligned} &\Delta u + K(|\mathbf{x}|)|u|^{p-2}u = 0, \\ &u \in C^2_{\mathrm{loc}}(\mathbb{R}^N), \end{aligned} \quad (27.49)$$

with $p > 1$. We assume that K and $rK'(r)$ belong to $C_{\mathrm{loc}}(\mathbb{R}^+)$.

THEOREM 27.24. *Under the above assumptions, for any classical radial solution to (27.49) the following is true:*

$$\frac{r^2}{p}K(r)|u(r)|^p + \frac{\alpha}{2}(N-\alpha-2)|u(r)|^2 + (N-\alpha-2)\int_0^r |u'(s)|^2\, s\, ds$$
$$+\int_0^r \left[\left(\alpha-\frac{2}{p}\right)K(s) - \frac{1}{p}sK'(s)\right]|u(s)|^p\, s\, ds$$
$$\leq \frac{\alpha}{2}(N-2)|u(0)|^2 \qquad \text{for } r>0. \quad (27.50)$$

Proof. From identity (27.47) we have for the solutions to equation (27.49) under consideration

$$\frac{1}{2}r^2|u'(r)|^2 + \frac{r^2}{p}K(r)|u(r)|^p + \alpha\, r\, u'(r)\, u(r) + \frac{\alpha}{2}(N-2)|u(r)|^2$$
$$+(N-\alpha-2)\int_0^r |u'(s)|^2\, s\, ds + \int_0^r \left[\left(\alpha-\frac{2}{p}\right)K(s) - \frac{1}{p}sK'(s)\right]|u(s)|^p\, s\, ds$$
$$= \frac{\alpha}{2}(N-2)|u(0)|^2 \qquad \text{for } r>0.$$

From this relation, by the inequality

$$-\frac{\alpha^2}{2}|u(r)|^2 \leq \frac{1}{2}r^2|u'(r)|^2 + \alpha\, r\, u'(r)\, u(r),$$

we obtain estimate (27.50). □

From this theorem we get immediately the following result.

THEOREM 27.25. *Let K and rK_r belong to $C_{\text{loc}}(\mathbb{R}^+)$ and K be such that there exists a constant α, $0 \leq \alpha \leq N-2$, such that*

$$(\alpha p - 2)\, K(r) - r\, K(r) \geq 0 \qquad \text{for } r>0.$$

Then, for each radial classical solution u to (27.49) with $N \geq 3$ the following estimate is true:

$$r^2 K(r)\, |u(r)|^p \leq \frac{\alpha\, p}{2}(N-2)\, |u(0)|^2 \qquad \text{for } r>0.$$

As an application of this inequality, consider equation (27.49) in $\mathbb{R}^3$.

COROLLARY. *Let K and rK_r belong to $C_{\text{loc}}(\mathbb{R}^+)$, $p>1$, and*

$$(p-2)K(r) - r\, K'(r) \geq 0 \qquad \text{for } r>0.$$

Then, for any radial classical solution u to (27.49) with $N=3$, the following estimate is true:

$$r^2 K(r)\, |u(r)|^p \leq \frac{p}{2}\, |u(0)|^2 \qquad \text{for } r>0. \quad (27.51)$$

27.30 SHARPNESS OF THE ESTIMATE

Estimate (27.51) can not be improved with respect to both the asymptotics as $r \to \infty$ and the constant on the right-hand side of this inequality. In order to verify this, consider the following example.

EXAMPLE. Let

$$\begin{aligned} &\Delta u + \omega^2 r^{p-2} |u|^{p-2} u = 0, \\ &u \in C^2_{\mathrm{loc}}(\mathbb{R}^3), \end{aligned} \tag{27.52}$$

with some constant $\omega \neq 0$ and $p \geq 2$. Then inequality (27.51) for classical radial solutions assumes the form

$$r^p \omega^2 \, |u(r)|^p \leq \frac{p}{2} \, |u(0)|^2 \qquad \text{for } r > 0.$$

Hence,

$$|u(r)| \leq \left[\frac{p\, u^2(0)}{2\omega^2} \right]^{1/p} r^{-1} \qquad \text{for } r > 0.$$

On the other hand, equation (27.52) has radial solutions of the form

$$u(r) = w(r)\, r^{-1}, \tag{27.53}$$

where w is a periodic nontrivial solution to the Cauchy problem

$$\begin{aligned} &w'' + \omega^2 |w|^{p-2} w = 0, \qquad r \in \mathbb{R}, \\ &w(0) = 0, \\ &w'(0) = A \neq 0. \end{aligned} \tag{27.54}$$

Hence, the decay estimate as $r \to \infty$ can not be improved.

Furthermore, for radial solutions (27.53) of equation (27.52) inequality (27.51) assumes the form

$$\omega^2 \, |w(r)|^p \leq \frac{p}{2} \, A^2 \qquad \text{for } r > 0. \tag{27.55}$$

This inequality in turn can not be improved. Indeed, for the periodic solutions to problem (27.54), the obvious identity

$$\frac{|w'(r)|^2}{2} + \frac{\omega^2 |w(r)|^p}{p} \equiv \frac{A^2}{2}$$

is true. Consequently, at the points $r_n > 0$, $n = 1, 2, \ldots$, where $w'(r_n) = 0$, inequality (27.55) becomes an equality.

Thus, inequality (27.51) is unimprovable, both with respect to the asymptotics as $r \to \infty$ and the constant on the right-hand side of the inequality.

As a particular case of equation (27.52) with $p = 2$, consider the Helmholtz equation

$$\begin{aligned} &\Delta u + \omega^2 u = 0 \qquad r \geq 0, \\ &u \in C^2(\mathbb{R}^3), \end{aligned} \tag{27.56}$$

where $\omega = \text{Const} \neq 0$. In this case inequality (27.51) takes the form

$$r^2\omega^2 |u(r)|^2 \leq |u(0)|^2 \qquad \text{for } r > 0.$$

It is easy to see that this inequality is sharp for radial solutions of the Helmholtz equation

$$u(r) = \frac{A}{\omega}\frac{\sin \omega r}{r} \qquad \text{for } r > 0$$

with $u(0) = A$, $A \neq 0$.

27.31 ADDITIONAL ESTIMATES

It is clear that, under additional conditions, the asymptotical estimate can be improved. To begin with, we establish the following result:

THEOREM 27.26. *Let K and rK_r belong to $C_{\text{loc}}(\mathbb{R}^+)$ and $p > 2$. Then for each classical radial solution to (27.49), the following inequality is satisfied:*

$$\begin{aligned} &\frac{1}{p} r^{k+1} K(r)|u(r)|^p + \frac{\alpha}{2}(N - \alpha - k - 1)r^{k-1}|u(r)|^2 \\ &\qquad + \left(N - \alpha - 1 - \frac{k+1}{2}\right) \int_0^r |u'(s)|^2 s^k\, ds \\ &\quad + \int_0^r \left[\left(\alpha - \frac{k+1}{p}\right) K(s) - \frac{1}{p} s K'(s)\right] |u(s)|^p s^k ds \\ &\qquad \leq \alpha \frac{(N-k-1)(k-1)}{2} \int_0^r |u(s)|^2 s^{k-2} ds, \qquad k > 1, \end{aligned} \tag{27.57}$$

for $r > 0$.

Proof. From identity (27.48) we have for the solutions of (27.49) under consideration

$$\frac{1}{2}r^{k+1}|u'(r)|^2 + \frac{r^{k+1}}{p}K(r)|u(r)|^p + \alpha\, r^k u'(r)\, u(r) + \frac{\alpha}{2}(N-k-1)r^{k-1}|u(r)|^2$$
$$+\left(N-\alpha-1-\frac{k+1}{2}\right)\int_0^r |u'(s)|^2\, s^k\, ds$$
$$+\int_0^r \left[\left(\alpha-\frac{k+1}{p}\right)K(s)-\frac{1}{p}sK'(s)\right]|u(s)|^p\, s^k ds$$
$$= \alpha\frac{(N-k-1)(k-1)}{2}\int_0^r |u(s)|^2\, s^{k-2} ds, \qquad k>1,$$

for $r>0$. □

From this relation and the inequality

$$-\frac{\alpha^2}{2}r^{k-1}|u(r)|^2 \le \frac{1}{2}r^{k+1}|u'(r)|^2 + \alpha\, r^k u'(r)\, u(r),$$

estimate (27.57) follows.

THEOREM 27.27. *Let K and rK_r belong to $C_{\rm loc}(\mathbb{R}^+)$, $K(r)>0$ for $r>0$, and let there exist constants $k>1$ and α,*

$$\alpha+\frac{k+1}{2}\le N+1,$$

such that

$$(\alpha\, p-k-1)\, K(r) - r\, K'(r) \ge 0 \qquad \text{for } r>0.$$

Let, for $p>2$,

$$\int_1^\infty ([s^{k+1}K(s)]^{-2/p} + [s^2K(s)]^{-2/(p-2)})s^{k-2}\, ds < \infty. \tag{27.58}$$

Then, for any radial classical solution u to (27.49) with $p>2$, the following estimate is satisfied:

$$r^{k+1}K(r)\,|u(r)|^p \le C_1 + C_2\, r^{k-(p+2)/(p-2)}[K(r)]^{-2/(p-2)} \qquad \text{for } r>0,$$

where positive constants C_1 and C_2 do not depend on r.

Proof. From the assumptions of the theorem inequality (27.57) implies

$$\frac{1}{p} r^{k+1} K(r)|u(r)|^p \le \alpha \frac{(N-k-1)(k-1)}{2} \int_0^r |u(s)|^2 s^{k-2} ds - \frac{\alpha}{2}(N-\alpha-k-1) r^{k-1}|u(r)|^2$$

for $r > 0$. Hence, we have for $p > 2$

$$\frac{1}{2p} r^{k+1} K(r)|u(r)|^p \le \alpha \frac{(N-k-1)(k-1)}{2} \int_0^r |u(s)|^2 s^{k-2} ds + C\, r^{k-(p+2)/(p-2)} [K(r)]^{-2/(p-2)}$$

for $r > 0$, where the positive constant C does not depend on r.

Then, denoting

$$Y(r) = \int_0^r |u(s)|^2 s^{k-2} ds,$$

we obtain the differential inequality

$$Y'(r) \le A_1 r^{k-2-2(k+1)/p} [K(r)]^{-2/p} [Y(r)]^{2/p} + A_2 r^{k-2-4/(p-2)} [K(r)]^{-2/(p-2)} \tag{27.59}$$

for $r > 0$, where the positive constants A_1 and A_2 do not depend on r. Hence, by the assumptions of the theorem, it follows that

$$Y(r) \le \lim_{r\to\infty} Y(r) < \infty.$$

Inequality (27.59) now implies the statement of the theorem. □

In the case of linear equation (27.49) with $p = 2$, the following result holds.

THEOREM 27.28. *Let all the assumptions of Theorem 27.27 with $p = 2$ be true, except inequality (27.58). Let there exist an $R > 0$ such that*

$$r^2 K(r) + \alpha (N - \alpha - k - 1) > 0 \qquad \text{for } r > R$$

and

$$\int_R^\infty \frac{dr}{[r^2 K(r) + \alpha (N-\alpha-k-1)]\, r} < \infty.$$

Then, for any classical radial solution u to equation (27.49) with $p = 2$ the inequality

$$r^{k-1} [r^2 K(r) + \alpha (N-\alpha-k-1)]\, |u(r)|^2 \le C_0 \tag{27.60}$$

is satisfied for $r > 0$, where the positive constant C_0 does not depend on r.

Remark. From (27.60) it follows that, under the conditions of the theorem, the best estimate of the decaying of a classical radial solution to (27.49) with $p = 2$ is obtained for the largest value of the exponent k (under the conditions of this theorem).

To clarify this remark, consider the well-known example of the Helmholtz equation

$$\Delta u + \omega^2 u = 0 \qquad \mathbf{x} \in \mathbb{R}^N, \qquad N \geq 4, \tag{27.61}$$

with a constant $\omega^2 > 0$.

Here the conditions of Theorem 27.28 reduce to

$$1 < \frac{k+1}{2} \leq \alpha \leq N - \frac{k+1}{2} - 1.$$

Hence, we find the largest $k = N - 2 > 1$ for $N \geq 4$.

Consequently, we get from inequality (27.60) the well-known sharp estimate of the decaying of classical radial solutions to (27.61).

$$r^{N-1}|u(r)|^2 < \text{Const} < \infty \qquad \text{for } r > 0,$$

where Const is independent of r.

§28. The Fibering Method. Existence of Infinitely Many Solutions

The fibering method is based on the representation of a solution of a corresponding equation in the form

$$u = tv. \tag{28.1}$$

Here t is a real parameter ($t \neq 0$ in some open set $I \subset \mathbb{R}$), and v is an element of the Banach space X ($v \neq 0$) satisfying the condition

$$H(t, v) = c \neq 0. \tag{28.2}$$

Generally speaking, any functional satisfying a sufficiently general condition can be taken as the fibering functional $H(t, v)$.

In particular, the norm $H(t, v) \equiv \|v\|$ can be taken as such a fibering functional $H(t, v)$. Then the condition (28.2) takes the form $\|v\| = 1$. In this case we get a so-called spherical fibering.

Here a solution $u \neq 0$ of a corresponding problem is sought in the form (28.1), where $t \in \mathbb{R}$ and $v \in S = \{w \in X, \|w\| = 1\}$.

Thus, the essence of the fibering method consists in embedding the space X of the original problem (28.1) in the larger space $\mathbb{R} \times X$ and investigating the new problem of conditional solvability in the new space $\mathbb{R} \times X$ under condition (28.2).

This method makes it possible to get both new solvability theorems and new theorems on the absence of a solution for nonlinear boundary value problems.

Further, in the investigation of solvability of boundary value problems this method makes it possible to separate the algebraic and the topological factors of the problem, which affect the number of its solutions.

28.1 General Theory of the Fibering Method in Variational Problems

Let X be a real Banach space with a norm $\|w\|_X$, that is, differentiable for $w \neq 0$, and let J be a functional on X of class $C^1(X \setminus \{0\})$.

We associate with the functional J a functional $\bar{J}$ defined on $\mathbb{R} \times X$ by

$$\bar{J}(t,v) = J(tv),$$

where $t \in \mathbb{R}$ and $v \in X$.

We consider this functional for $(t,v) \in I \times S$. Here I is an arbitrary nonempty open set in $\mathbb{R}$, and S is the unit sphere in $X : S = \{u \in X, \|u\| = 1\}$.

THEOREM 28.1. *Let X be a Banach space with a norm differentiable on $X \setminus \{0\}$, and let $(t,v) \in (I \setminus \{0\}) \times S$ be a conditionally critical point of the functional $\bar{J}$, regarded on $I \times S$. Then the vector $u = tv$, $t \neq 0$, is a critical point of the functional J, i.e., $J'(u) = 0$.*

Proof. The proof is obvious. Indeed, at the conditionally critical point $(t,v) \in (I \setminus \{0\}) \times S$ we have that

$$\lambda \bar{J}'_v(t,v) = \mu \|v\|', \tag{28.3}$$

$$\bar{J}'_t(t,v) = 0, \tag{28.4}$$

with $\lambda^2 + \mu^2 \neq 0$.

By (28.3),

$$\lambda \left\langle \bar{J}'_v(t,v), v \right\rangle = \mu \left\langle \|v\|, v \right\rangle,$$

where $\langle w^*, u \rangle$ is the value of a functional w^* in the dual space X^* on an element u in X. Then from this and the equalities

$$\left\langle \bar{J}'_v(t,v), v \right\rangle = t \left\langle J'(tv), v \right\rangle = t\bar{J}'_t \left\langle t, v \right\rangle,$$

$$\left\langle \|v\|', v \right\rangle = 1 \qquad \text{for } v \in S,$$

we get that $\lambda t \bar{J}(t,v) = \mu$.

It follows from this equality and (28.4) that $\mu = 0$. Then $\lambda \neq 0$, and by (28.3)

$$tJ'(u) = \bar{J}'_v(t,v) = 0$$

for $u = tv$, $t \neq 0$. Consequently, $J'(u) = 0$, and Theorem 28.1 is proved. □

Now we consider a more general fibering (decomposition).

For this we introduce a fibering functional $H(t,v)$ defined on $\mathbb{R} \times X$ and we consider the functional $\bar{J}(t,v)$ under the condition (28.2). As $H(t,v)$ we can in general take an arbitrary functional that is differentiable under condition (28.2) and satisfies the inequality

$$\langle H_v, v \rangle \neq tH_t \qquad \text{for } H(t,v) = c \neq 0. \tag{28.5}$$

THEOREM 28.2. *Let H be a functional of the indicated class. Let $(t, v) \in I \times X$ with $t \neq 0$ and $v \neq 0$ be a conditionally critical point of the functional $\bar{J}(t, v)$ under condition (28.2). Then the point $u = tv$ is a nonzero critical point of the original functional J, i.e., $J'(u) = 0$ and $u \neq 0$.*

Proof. The proof is analogous to the preceding proof. Indeed, at the conditionally critical point (t, v) we have that

$$\mu \bar{J}'_v(t, v) = \lambda H'_v(t, v), \qquad \mu \bar{J}'_t(t, v) = \lambda H'_t(t, v), \tag{28.6}$$

with $\mu^2 + \lambda^2 \neq 0$; on the other hand,

$$\bar{J}'_v(t, v) = tJ'_u(tv), \qquad \bar{J}'_t(t, v) = \left\langle J'_\mu(tv), v \right\rangle.$$

Then from (28.6) we get that

$$\mu t J'_u(tv) = \lambda H'_v(t, v), \qquad \mu \left\langle J'_u(tv), v \right\rangle = \lambda H'_t(t, v). \tag{28.7}$$

From this we get

$$\mu t \left\langle J'_u(t, v), v \right\rangle = \lambda \left\langle H'_v(t, v), v \right\rangle,$$
$$\mu t \left\langle J'_u(tv), v \right\rangle = \lambda t H'_t(t, v).$$

Consequently,

$$\lambda \left\langle H'_v(t, v), v \right\rangle = \lambda t H'_t(t, v)$$

for $t \neq 0$ and $H(t, v) = c \neq 0$. Then, by condition (28.5), we get that $\lambda = 0$, and hence $\mu \neq 0$.

As a result, the first equation in (28.7) takes the form $J'(u) = 0$, with $u = tv \neq 0$. □

Remark. Obviously, in this theorem the condition of differentiability of the norm $\|w\|$ in X for $w \neq 0$ is superfluous.

28.2 MINIMAX REALIZATION OF THE FIBERING METHOD IN VARIATIONAL PROBLEMS

Let X be a real Banach space with a norm differentiable on $X \setminus \{0\}$, let J be a functional on X belonging to the class $C^1(X \setminus \{0\})$, and let I be a nonempty open subset of $\mathbb{R}$. Now we consider the functional

$$\sup_{t \in I} J(tv)$$

for $v \in S$.

THEOREM 28.3. *Suppose that for any $v \in S$ the quantity*

$$\bar{J}(v) = \max_{t\in I} J(tv) \tag{28.8}$$

exists, and $J(v) > J(0)$ if $0 \in I$. Assume that f is differentiable on the sphere S. Then to each conditionally critical point v_c of the functional J, regarded on the unit sphere S, there corresponds a critical point $u_c = t_c v_c$ of J with $t_c \in I \setminus \{0\}$ such that $J(t_c v_c) = \bar{J}(v_c)$.

Proof. Assume the theorem is false, i.e., $J'(u_c) \neq 0$, and hence $t_c J'(u_c) \neq 0$, because $t_c \neq 0$. Then there exists an element $w_0 \in X$ such that

$$t_c \langle J'(u_c), w_0\rangle > 0. \tag{28.9}$$

Because $v_c \in S$ is a conditionally critical point of the functional J, which is differentiable on S, it follows that

$$\bar{J}\left(\frac{v_c + \xi w_0}{\|v_c + \xi w_0\|}\right) = \bar{J}(v_c) + \varepsilon(\xi)\xi = J(t_c v_c) + \varepsilon(\xi)\xi \tag{28.10}$$

for sufficiently small ξ, with $\varepsilon(\xi) \to 0$ as $\xi \to 0$.

On the other hand, by the definition of the functional $\bar{J}$,

$$J\left(t\frac{v_c + \xi w_0}{\|v_c + \xi w_0\|}\right) \leq \bar{J}\left(\frac{v_c + \xi w_0}{\|v_c + \xi w_0\|}\right) \qquad \text{for all } t \in I. \tag{28.11}$$

By a condition of the theorem, $\max_{t\in I} J(tv_c) = J(t_c v_c)$ is attained on the open set $I \setminus \{0\}$; hence $t_c\|v_c + \xi w_0\| \in I \setminus \{0\}$ for sufficiently small ξ, because $\|v_c\| = 1$. Then, setting $t = t_c\|v_c + \xi w_0\|$ in (28.11) for sufficiently small ξ, we get by (28.10) that

$$J(t_c v_c + \xi t_c w_0) \leq J(t_c v_c) + \varepsilon(\xi)\xi. \tag{28.12}$$

Because J is differentiable,

$$J(t_c v_c + \xi t_c w_0) = J(t_c v_c) + \xi t_c \langle J'(u_c), w_0\rangle + \varepsilon_1(\xi)\xi,$$
$$\varepsilon_1(\xi) \to 0 \qquad \text{as } \xi \to 0.$$

Then it follows from (28.10) that

$$\xi t_c \langle J'(u_c), w_0\rangle \leq \xi\varepsilon_2(\xi),$$
$$\varepsilon_2(\xi) \to 0 \qquad \text{as } \xi \to 0.$$

Considering this inequality for sufficiently small positive ξ, we get

$$t_c \langle J'(u_c), w_0\rangle \leq 0.$$

In view of (28.9) this inequality contradicts the assumption. Theorem 28.3 is proved. □

Remark. Let I be a nonempty open set in $\mathbb{R}$ that is symmetric with respect to $0 \in \mathbb{R}$, i.e., $t \in I$ implies that $-t \in I$.

Suppose that the functional $\bar{J}$ defined by (28.8) exists. Then it is even. This makes it possible to use the Lyusternik-Shnirel'man theory for certain functionals that are not even, and obtain theorems on the existence of many geometrically different critical points.

28.3 On the Connection Between Conditionally Critical Points and Critical Points of Functionals

In the preceding sections the fibering method was used to establish a connection between conditionally critical points and critical points of functionals. We now consider this connection from a somewhat different point of view.

Let η be a differentiable mapping (which can also be constant) from a Banach space X into X. With a functional J twice differentiable on X we associate a functional J_η by the formula

$$J_\eta(u) = \langle J'(u), \eta(u) \rangle \, .$$

It is clear that a critical point u of J satisfies the condition $J_\eta(u) = 0$ and hence is a conditionally critical point of J, considered under the condition $J_\eta(u) = 0$. The following simple theorem contains conditions for the validity of the converse.

Theorem 28.4. *Let J be a twice differentiable functional on a Banach space X. Suppose that there exists a differentiable mapping η from X to X such that for a conditionally critical point u_0 of the functional J, considered under the condition $J_\eta(u) = 0$, the inequality*

$$\langle J'_\eta(u_0), \eta(u_0) \rangle \neq 0 \tag{28.13}$$

holds for $J_\eta(u_0) = 0$. Then the conditionally critical point u_0 of J is an unconditionally critical point of J.

Proof. The proof is obvious. Indeed, at u_0

$$\lambda J'(u_0) = \mu J'_\eta(u_0), \qquad \lambda^2 + \mu^2 \neq 0. \tag{28.14}$$

Hence,

$$\lambda \langle J'(u_0), \eta(u_0) \rangle = \mu \langle J'_\eta(u_0), \eta(u_0) \rangle \qquad \text{with } \langle J'(u_0), \eta(u_0) \rangle = J_\eta(u_0) = 0.$$

Using (28.13), we get that $\mu = 0$, and hence $\lambda \neq 0$. Then (28.14) implies that $J'(u_0) = 0$. Theorem 28.4 is proved. □

28.4 THE CHOICE OF THE FIBERING FUNCTIONAL

A realization of the fibering method depends on the choice of the fibering functional $H(t,v)$ satisfying condition (28.5).

As the simplest of such functionals we can take the norm of the Banach space under the condition that it is differentiable away from the zero of the space. However, this choice is not unique.

Here we propose a fibering functional naturally generated by the problem itself, i.e., the functional J. We remark, however, that this choice is not necessary.

The scalar equation

$$\langle J'(tv), v\rangle = 0 \tag{28.15}$$

in the scalar parameter $t = t(v)$ is the defining equation in the fibering method.

To separate algebraically different solutions of this equation it seems natural to take the following as a fibering functional H in the case of a functional J of class $C^3(X \setminus \{0\})$:

$$H(t,v) \equiv \langle J''(tv)v, v\rangle \,. \tag{28.16}$$

The standard notation is used here and below.

In this case the relation

$$\langle H_v, v\rangle - tH_t = 2H \tag{28.17}$$

is valid for the functional H.

The proof uses the following equalities:

$$\langle H_v, v\rangle = \frac{d}{d\zeta} H(t,\zeta v)|_{\zeta=1}$$

and

$$tH_t = \frac{d}{d\zeta} H(t\zeta, v)|_{\zeta=1}.$$

Then

$$H(t,\zeta v) = \zeta^2 \langle J''(\zeta tv)v, v\rangle$$

and

$$\langle H_v, v\rangle = \frac{d}{d\zeta} H(t,\zeta v)|_{\zeta=1} = 2\langle J''(tv)v, v\rangle + \frac{d}{d\zeta}\langle J''(\zeta tv)v, v\rangle\,|_{\zeta=1} = 2H + tH_t.$$

By (28.17), the condition (28.5) for nondegeneracy of the fibering functional H defined by (28.16) turns out to hold always. When such a fibering functional is chosen, the solution $t = t(v)$ of (28.15) a priori inherits the smoothness (with a loss of one derivative) of the original functional on this set, if this solution exists.

Accordingly, the problem of finding critical points (different from the zero of X) of the functional $J \in C^3(X \setminus \{0\})$ reduces to the problem of finding the conditionally critical points of the functional $J(tv)$ under the condition

$$\langle J''(tv)v, v\rangle = c \neq 0, \qquad t \in I.$$

The substitution $v \to v_1 = kv$ $(k \neq 0)$ reduces this problem to the problem of finding the conditionally critical points of $J(tv)$ under the condition

$$\langle J''(tv)v, v\rangle = c_0, \qquad \text{and } t \in I_1. \tag{28.18}$$

Here the scalar parameter c_0 takes either the value $+1$ or the value -1, and I_1 is the corresponding image $(t \to t_1 = t/k)$ of the original open set I. In this problem the previous notation t and v is kept instead of t_1 and v_1.

THEOREM 28.5. *Let J be a functional on X in the class $C^3(X \setminus \{0\})$, and let $(t, v) \in I_1 \times X$ with $t \neq 0$ and $v \neq 0$ be a conditionally critical point of the functional $J(tv)$ under condition (28.18).*

Then the point $u = tv$ is a nonzero critical point of J.

The proof follows immediately from Theorem 28.2.

28.5 A PARAMETER-FREE REALIZATION OF THE FIBERING METHOD

In the general case the fibering method reduces the original variational problem to a parametric variational problem and to the investigation of its conditionally critical points.

In the case when the fibering functional H of the original variational problem is defined by (28.17) it is possible to eliminate the parameter t in the new variational problem. Indeed, condition (28.18) for a functional J of class $C^3(X \setminus \{0\})$ means that

$$\frac{d}{dt}\langle J'(tv), v\rangle = \langle J''(tv)v, v\rangle \neq 0$$

on the set (28.18).

Thus, condition (28.18) for a functional J of class $C^3(X \setminus \{0\})$ enables us to single out in the defining equation (28.15) the algebraically different smooth solutions $t_i(v)$ $(i = 1, \dots, N)$ when they exist, and to eliminate the scalar parameter t.

Then the problem of finding the critical points (different from the zero of X) of a functional $J \in C^3(X \setminus \{0\})$ reduces to the problem of finding the conditionally critical points of the functional

$$\begin{cases} \mathcal{F}_i(v) = J(t_i(v)v) \\ \text{under the corresponding conditions} \\ H_i(v) \equiv \langle J''(t_i(v))v, v\rangle = c_0 \end{cases} \tag{28.19}$$

with $c_0 = \pm 1$.

Here $t_i(v)$ is the corresponding solution of (28.15) $(i = 1, \dots, N)$.

THEOREM 28.6. *Let J be a functional defined on X and in the class $C^3(X \setminus \{0\})$, let $t_i(v)$ be the solution of (28.15) under condition (28.18), and let v_i be* *a conditionally critical point of problem (28.19) with* $t_i(v_i) \neq 0$. *Then the point* $u_i = t_i(v_i)v_i$ *is a nonzero critical point of* J.

Proof. The proof follows from Theorem 28.5, because the pair $(t_i, v_i) \in I_1 \times X$ satisfies the condition of this theorem. □

28.6 A Particular Realization of the Fibering Method

From the functional point of view it is sometimes convenient in studying a variational problem to take for the fibering (normalizing) functional H the principal part of the original functional of norm type. When there is a complementary weakly continuous functional, the original problem reduces to the problem of investigating conditionally critical points of a continuous functional on a sphere-type surface. It is simpler to investigate a variational problem in a closed ball, because the latter is convex.

We present a class of variational problems in which this approach can be implemented.

Suppose that on a Banach space X with a norm differentiable away from zero the given functional J of class $C^1(X)$ has the form $J(u) = J_0(u) + J_1(u)$, where $J_0(u) = \|u\|^p$, for some $p > 1$ and that $J_1 \in C^1(X)$. Here we choose the fibering functional H to be $H(v) = \|v\|^p$, so that condition (28.2) takes the form $\|v\| = 1$, i.e., we use the method of spherical fibering.

The functional $J(tv)$ takes the form

$$J^*(t,v) = |t|^p + J_1(t,v) \qquad \text{for } v \in S,$$

and the equation

$$J_0'(u) + J_1'(u) = 0 \tag{28.20}$$

is then equivalent to the system

$$\begin{aligned} p\,|t|^{p-2}t + \langle J_1'(tv), v\rangle &= 0, \qquad t \neq 0, \\ t\,J_1'(tv) &= \nu\,\|v\|', \qquad v \in S. \end{aligned} \tag{28.21}$$

Here the prime denotes the derivative of the corresponding functional.

Note that here the Lagrange multiplier ν differs from the corresponding multiplier λ in Section 28.3, where another realization of the fibering method was considered.

The functional $\langle J_1'(tv), v\rangle$ is defined for all $v \in X$, so the first scalar equation (28.21) in t can be considered for $v \in B = \{w \in X, \|w\| \leq 1\}$. Suppose that this equation has solutions $t_i(w)$ $(i = 1, \ldots, N)$. Let

$$\mathcal{F}_i(w) = |t_i|^p + J_1(t_i(w)w)$$

and consider these functionals on the closed unit ball B.

DEFINITION. A point $w \in B$ is called a *critical point of the differentiable functional* $\mathcal{F}_i$ in the closed unit ball B if one of the following two conditions holds: either w lies in the interior of B and is an ordinary critical point of $\mathcal{F}_i$; or w lies on the boundary $\partial B = S$ and is a conditionally critical point of the functional $\mathcal{F}$ on the sphere S.

DEFINITION. A critical point $w \in B$ of the differentiable functional $\mathcal{F}_i$ is said to be a regular critical point of it if $w_i \neq 0$, $t_i(w_i) \neq 0$, and the functional t_i is differentiable at w_i. Here $t_i(w)$ is a solution of (28.21) for $v = w \in B$.

THEOREM 28.7. *Suppose that the functional $\mathcal{F}_i$ has a regular critical point $w_i \in B$. Then $w_i \in S$, and $u_i = t_i(w_i)\, w_i$ is a nonzero solution of (28.20).*

Proof. Suppose that the regular critical point $w_i \in B$ of $\mathcal{F}_i$ is in the interior of the unit ball B.

We study the behavior of $\mathcal{F}_i$ along the ray ζw_i as $\zeta \to 1$. By the differentiability of J and the differentiability of the functional t_i at the point w_i,

$$\left.\frac{d\mathcal{F}_i(\zeta\, w_i)}{d\zeta}\right|_{\zeta=1} = (p\,|t_i|^{p-2}t_i + \langle J_1'(t_i w_i), w_i\rangle)\left.\frac{d\,t_i}{d\zeta}\right|_{\zeta=1} + \langle J_1'(t_i w_i), w_i\rangle\; t_i.$$

Here $t_i = t(w_i)$. Hence, in view of (28.21) with $v = w_i$,

$$\left.\frac{d\mathcal{F}_i(\zeta w_i)}{d\zeta}\right|_{\zeta=1} = t_i\;\langle J_1'(t_i w_i), w_i\rangle = -p\,|t_i|^p \neq 0.$$

Thus, a regular critical point w_i of $\mathcal{F}_i$ belongs to the boundary $\partial B = S$.

Then by Theorem 28.1 the point $u_i = t_i(w_i)\, w_i$ is a nonzero solution of (28.20). □

Remark. Following the above definitions, we introduce the concept of a regular extremal point $w \in B$ of the functional $\mathcal{F}_i(w)$ by replacing the word "critical" in those definitions by the word "extremal".

As a corollary to Theorem 28.7 we get the following "maximum principle".

COROLLARY. *Suppose that the functional $\mathcal{F}_i(w)$ has a regular extremal point $w_i \in B$. Then $w_i \in \partial B$, and $u_i = t_i(w_i)w_i$ is a nonzero solution of (28.20).*

28.7 A SIMPLE EXAMPLE

Now we demonstrate the use of the method of spherical fibering in simple examples of known problems.

In these examples the defining equation (28.14) with $v \in S$ admits an explicit smooth solution $t = t(v)$ for $v \in S$. This makes it possible to use a parameter-free realization of the method of spherical fibering (see Sections 28.5 and 28.6).

As a simple example consider the linear Dirichlet problem

$$\Delta u - u = h(\mathbf{x}) \qquad \text{in } \mathbb{R}^N \tag{28.22}$$

with nontrivial $h \in \mathcal{E}_2^*$. Corresponding to this problem is the functional

$$\Phi(u) = -\frac{1}{2}\|\nabla u\|_2^2 - \frac{1}{2}\|u\|_2^2 - \int_{\mathbb{R}^N} hu\, d\mathbf{x}.$$

According to the fibering method, we set $u = tv$.

Then the functional Φ takes the form

$$\Phi(tv) = -\frac{1}{2}t^2\|\nabla u\|_2^2 - \frac{1}{2}t^2\|u\|_2^2 - t\int_{\mathbb{R}^N} hu\, d\mathbf{x}.$$

In the spherical fibering

$$\|v\|_{\mathcal{E}_2}^2 = \|\nabla u\|_2^2 + \|u\|_2^2 = 1$$

the functional

$$\Phi^*(t,v) = -\frac{1}{2}t^2 - t\int_{\mathbb{R}^N} hv\, d\mathbf{x} \tag{28.23}$$

then corresponds to the functional $\Phi(tv)$. From the defining equation

$$\Phi_t^*(t,v) = -t - \int_{\mathbb{R}^N} hv\, d\mathbf{x} = 0$$

we find that

$$t = -\int_{\mathbb{R}^N} hv\, d\mathbf{x}$$

and we substitute it in the expression (28.23) for Φ^*. Then

$$\mathcal{F}(v) = \Phi^*(t(v),v) = \frac{1}{2}\left(\int_{\mathbb{R}^N} hv\, d\mathbf{x}\right)^2.$$

Note that the minimax realization of the fibering method

$$\Phi^*(t,v) \to \max_{t\in\mathbb{R}} \Phi^*(t,v) = \frac{1}{2}\left(\int_{\mathbb{R}^N} hv\, d\mathbf{x}\right)^2$$

gives rise to the very same functional $\mathcal{F}$.

We now consider the critical points of this even functional on the unit sphere. Obviously, there exists an infinite set of conditionally critical points of $\mathcal{F}$ on the unit sphere. In this set there are only two regular conditionally critical points, v_1 and $v_2 = -v_1$, i.e., such that (in this case) $t_1 = t(v_2) \neq 0$ and $t_2 = t(v_2) \neq 0$. These are points at which $\mathcal{F}(v)$ attains a maximum on the unit sphere.

We remark that v_1 and v_2 are clearly such maximum points of $\mathcal{F}$ on the closed unit ball $B = \{v \in \mathcal{E}_2 : \|\nabla v\|_2^2 + \|u\|_2^2 \leq 1\}$.

Then by Theorem 28.5, and also by Theorem 28.6, $u_1 = t_1 v_1$ and $u_2 = t_2 v_2$ are solutions of the Dirichlet problem (28.22), i.e.,

$$u_1 = -v_1 \int_{\mathbb{R}^N} h v_1 \, d\mathbf{x} \qquad \text{and } u_2 = -v_2 \int_{\mathbb{R}^N} h v_2 \, d\mathbf{x} = -v_1 \int_{\mathbb{R}^N} h v_1 \, d\mathbf{x} = u_1.$$

Thus, the two nonzero solutions coincide.

Remark. To clear up the essence of the fibering method we verify Theorems 28.5 and 28.6 directly in this example.

Thus, $v_1 \in S$ is a maximum point of $\mathcal{F}$ on the unit sphere S. Then by the Lagrange rule, at this point

$$h \int_{\mathbb{R}^N} h v_1 \, d\mathbf{x} = -\nu(\Delta v_1 - v_1), \qquad v_1 \in \mathcal{E}_2.$$

From this we find for $\|v_1\|_{\mathcal{E}_2} = 1$ that $\nu = \left(\int_{\mathbb{R}^N} h v_1 \, d\mathbf{x}\right)^2$, and $\nu \neq 0$, because $\max_{v \in S} \Phi(v) > 0$ for $\|h\|_{\mathcal{E}_2^*} \neq 0$. Then

$$h \int_{\mathbb{R}^N} h v_1 \, d\mathbf{x} = -\left(\int_{\mathbb{R}^N} h v_1 \, d\mathbf{x}\right)^2 (\Delta v_1 - v_1)$$

or, setting $t_1 = -\int_{\mathbb{R}^N} h v_1 \, d\mathbf{x} \neq 0$ and $u_1 = t_1 v_1$,

$$\Delta u_1 - u_1 = h(\mathbf{x}), \qquad u_1 \in \mathcal{E}_2,$$

i.e., u_1 is a solution of the problem (28.22). We see similarly that u_2 is a solution of this problem, and that $u_1 = u_2$.

Remark. We note that here the general formula (28.16) for the fibering functional $H(t, v)$ leads also to spherical fibering. Indeed,

$$H(t, v) \equiv \frac{d^2}{dt^2} \Phi(tv) = -\left(\|\nabla u\|_2^2 + \|u\|_2^2\right),$$

and general condition (28.18) takes the form

$$\|\nabla u\|_2^2 + \|u\|_2^2 = 1.$$

28.8 The Second Example. Infinitely Many Solutions

We do not include in this book questions connected with the existence of multiple solutions. Nevertheless we here consider an example that demonstrates the combined use of the fibering method and the Lyusternik-Shnirel'man theory. Reader who are unfamiliar with this theory may omit the present section or look up Krasnoselskii [1] or Berestycki and Lions [2].

Consider the problem

$$-\Delta u + a(\mathbf{x})u + b(\mathbf{x})|u|^{q-2} = 0 \qquad \text{in } \mathbb{R}^N \tag{28.24}$$

Here $2 < q < 2^*$. Let $a > 0$ and $b > 0$ satisfy the conditions of one of Theorems 7.8, 7.9, or 7.10.

The original functional Φ defined on $\mathcal{E}_{2(a),p(b)}$ or, by embedding, on $\mathcal{E}_{2(a)}$ has the form

$$\Phi(u) = \frac{1}{2}\|\nabla u\|_2^2 + \frac{1}{2}\|a^{1/2}u\|_2^2 - \frac{1}{q}\|b^{1/q}u\|_q^q .$$

According to the fibering method, we set $u = tv$. Then

$$\Phi(tv) = \frac{1}{2}t^2\|\nabla v\|_2^2 + \frac{1}{2}t^2\|a^{1/2}v\|_2^2 - \frac{|t|^q}{q}\|b^{1/q}v\|_q^q.$$

In the spherical fibering

$$\|v\|_{\mathcal{E}_{2(a)}} = \left(\|\nabla v\|_2^2 + \|a^{1/2}v\|_2^2\right)^{1/2} = 1$$

the functional

$$\Phi^*(t,v) = \frac{1}{2}t^2 - \frac{|t|^q}{q}\|b^{1/q}v\|_q^q$$

corresponds to the functional Φ. From the defining the bifurcation equation

$$\Phi_t^*(t,v) = t - |t|^{q-2}t\,\|b^{1/q}v\|_q^q = 0$$

we find explicitly the real nonzero solutions

$$t = \pm\left(\|b^{1/q}v\|_q^q\right)^{1/(2-q)}.$$

Then the functional $\mathcal{F}(v) = \Phi^*(t(v),v)$ takes the form

$$\mathcal{F}(v) = \frac{q-2}{2q}(\|b^{1/q}v\|_q)^{-2q/(q-2)}.$$

We supposed that a and b are such that $\mathcal{E}_{2(a)} \circlearrowleft\circlearrowleft L_{q(b)}$. Then to this functional, regarded on the unit sphere $S \subset \mathcal{E}_{2(a)}$, we can apply the Lyusternik-Shnirel'man theory, in view of which Φ has a countable set of geometrically different conditionally critical points $v_1, v_2, \ldots, v_m$ on S, with $\|v_m\|_{\mathcal{E}_{2(a)}} = 1$ and $\|b^{1/q}v_m\|_q \to 0$ as $m \to \infty$. Then by Theorem 28.6, and also by Theorem 28.7, we get that problem (28.23) has a countable set of geometrically different solutions $\pm u_1, \pm u_2, \ldots$ $\pm u_m, \ldots$ with

$$u_m(\mathbf{x}) = v_m(\mathbf{x})\|b^{1/q}v_m\|_q^{-q/(q-2)}$$

and $\|u_m\|_{\mathcal{E}_{2(a)}} \to \infty$ as $m \to \infty$.

Thus the following holds:

THEOREM 28.8. *Let a and b satisfy the conditions of one of Theorems 7.8, 7.9, or 7.10. Then problem (28.24) has infinitely many solutions $\pm u_m \in \mathcal{E}_{2(a)}$ with $\|u_m\|_{\mathcal{E}_{2(a)}} \to \infty$ as $m \to \infty$.*

§29. Nonexistence Results

The absence of solutions to elliptic equations on $\mathbb{R}^N$ in various classes is of a special interest. We consider two factors of nonexistence: algebraic and functional. These two factors are based on identities that we now derive.

29.1 An Algebraic Factor

On a Banach space X with norm differentiable away from zero, let a functional J of the class $C^1(X)$ be given in the form

$$J(u) = J_0(u) + J_1(u), \qquad J_0(u) = \|u\|^p, \qquad p > 1.$$

Then, as follows from §28, the existence of a nonzero critical point of this functional is equivalent to the existence of a solution (t, v) with $t \in \mathbb{R} \setminus \{0\}$ and $v \in S$ for the following system of equations:

$$\begin{cases} p\,|t|^{p-2}t + \langle J_1'(tv), v\rangle = 0, & t \neq 0, \\ |t|^{p-2}t\,(\|v\|^p)' + J_1'(tv) = 0, & v \in S. \end{cases}$$

Suppose now that w is some element of X. Then from this system we get the following system of two scalar equations:

$$\begin{cases} p\,|t|^{p-2}t + \langle J_1'(tv), v\rangle = 0, \\ |t|^{p-2}t\,\langle(\|v\|^p)', w\rangle + \langle J_1'(tv), w\rangle = 0. \end{cases} \tag{29.1}$$

From this system we derive the following test for the absence of a nonzero solution in X of the equation

$$J_0'(u) + J_1'(u) = 0. \tag{29.2}$$

THEOREM 29.1. *Let J_0 and J_1 be the functionals defined above, and suppose that there exists an element $w \in X$ such that system (29.1) is inconsistent whatever the values of $t \in \mathbb{R} \setminus \{0\}$ and $v \in S$. Then equation (29.2) does not have a nontrivial solution in X.*

We consider a simple example.

LEMMA 29.2. *Let $f : \mathbb{R}^N \times \mathbb{R} \to \mathbb{R}$ be a Carathéodory function. Let u be a generalized solution of the problem*

$$\begin{aligned} &-\Delta u + f(\mathbf{x}, u) = 0 \qquad \text{on } \mathbb{R}^N, \\ &u \to 0 \qquad \text{as } |\mathbf{x}| \to \infty, \end{aligned} \tag{29.3}$$

with $u\, f(\,.\,, u) \in L_1(\mathbb{R}^N)$. Then u satisfies the identity

$$\int_{\mathbb{R}^N} |\nabla u|^2\, d\mathbf{x} + \int_{\mathbb{R}^N} u\, f(\mathbf{x}, u)\, d\mathbf{x} = 0. \tag{29.4}$$

Proof. In accordance with the definition of a generalized solution in Section 1.4 we have $\nabla u \in L_2(\mathbb{R}^N)$ and $f(\mathbf{x},\,.\,) \in L_1(\mathbb{R}^N)$. Then

$$\int_{\mathbb{R}^N} (\nabla u, \nabla \varphi_n)\, d\mathbf{x} + \int_{\mathbb{R}^N} f(\mathbf{x}, u)\varphi_n\, d\mathbf{x} = 0 \tag{29.5}$$

for any $\varphi_n \in \mathcal{D}(\mathbb{R}^N)$. Choose the sequence $\{\varphi_n\}$ in such a way that

$$\begin{aligned} \nabla \varphi_n \to \nabla u \qquad &\text{in } L_2(\mathbb{R}^N), \\ f(\,.\,, u)\, \varphi_n \to f(\,.\,, u)\, u \qquad &\text{in } L_1(\mathbb{R}^N). \end{aligned}$$

Passing to the limit in (29.5) as $n \to \infty$ we get (29.4). □

THEOREM 29.3. *Let the conditions of Lemma 29.2 hold, $f \in K_{p,q}$, and $t\, f(\mathbf{x}, t) \geq 0$ for almost all $\mathbf{x} \in \mathbb{R}^N$ and all $t \in \mathbb{R}$. Then problem (29.3) has no nontrivial solution from $\mathcal{E}_{p,q}$.*

As a substantial example consider the problem

$$-\Delta u + \lambda_1 |u|^{p-2} u + \lambda_2 |u|^{q-2} u = 0, \qquad \text{in } \mathbb{R}^N. \tag{29.6}$$

Clearly, for $1 < p \leq q < +\infty$, $\lambda_1 \geq 0$, and $\lambda_2 \geq 0$ problem (29.3) has no nontrivial solutions from $\mathcal{E}_{p,q}$. Indeed, identity (29.5) assumes in this case the form

$$\|\nabla u\|_2^2 = -\lambda_1 \|u\|_p^p - \lambda_2 \|u\|_q^q,$$

which makes the statement obvious.

29.2 THE VARIATIONAL IDENTITY

In this section an identity permitting us to get finer results is derived. It is a particular case of the Pohozaev identity [2].

THEOREM 29.4. *Let $f, fx : \mathbb{R}^N \times \mathbb{R} \to \mathbb{R}$ be Carathéodory functions. Let $u \in C^2_{\text{loc}}(\mathbb{R}^N)$ be a solution of (29.3) such that $\nabla u \in L_2(\mathbb{R}^N)$, and $F(\,.\,, u)F_1(\,.\,, u) \in L_1(\mathbb{R}^N)$, where*

$$F(\mathbf{x}, t) = \int_0^t f(\mathbf{x}, s)\, ds$$

$$F_1(\mathbf{x}, t) = \sum x_i \frac{\partial F(\mathbf{x}, u)}{\partial x_i}.$$

Then

$$\frac{N-2}{2} \int_{\mathbb{R}^N} |\nabla u|^2 \, d\mathbf{x} + \int_{\mathbb{R}^N} \left(N\, F(\mathbf{x}, u) + \sum_{i=1}^N x_i \frac{\partial F(\mathbf{x}, u)}{\partial x_i} \right) d\mathbf{x} = 0. \tag{29.7}$$

Proof. Multiply equation (29.3) by $\sum_{i=1}^N x_i \nabla_i u$ and integrate over the ball B_R. Integrating by parts the multiplied second term of (29.3) gives

$$\int_{B_R} \sum_{i=1}^N x_i \nabla_i u\, f(\mathbf{x}, u)\, d\mathbf{x} = \int_{\partial B_R} F(\mathbf{x}, u) \sum_{i=1}^N x_i n_i \, d\sigma$$
$$- \int_{B_R} \left(N\, F(\mathbf{x}, u) + \sum_{i=1}^N x_i \frac{\partial F(\mathbf{x}, u)}{\partial x_i} \right) d\mathbf{x}, \tag{29.8}$$

where n_i are components of the unit outward normal to ∂B_R and $d\sigma$ is an area element. Integrating by parts the multiplied first term in (29.3), we get

$$- \int_{B_R} \sum_{j=1}^N \nabla_j \nabla_j u \sum_{i=1}^N x_i \nabla_i u \, d\mathbf{x}$$
$$= \int_{B_R} \sum_{j=1}^N \sum_{i=1}^N (\nabla_j u)\, (\delta_{ij} \nabla_j u + x_i \nabla_j \nabla_i u)\, d\mathbf{x} - \int_{\partial B_R} \sum_{j=1}^N \sum_{j=i}^N x_i n_j (\nabla_j u) \nabla_i u \, d\sigma, \tag{29.9}$$

where δ_{ij} is the Kronecker sign: $\delta_{ij} = 1$ if $i = j$, otherwise $\delta_{ij} = 0$. Integrating by parts provides

$$\int_{B_R} \sum_{j=1}^N \sum_{i=1}^N x_i (\nabla_j u) \nabla_i \nabla_j u \, d\mathbf{x} = \frac{1}{2} \int_{B_R} \sum_{i=1}^N x_i \nabla_i \sum_{j=1}^N |\nabla_j u|^2 \, d\mathbf{x}$$
$$= -\frac{N}{2} \int_{B_R} \sum_{j=1}^N (\nabla_j u)^2 \, d\mathbf{x} + \frac{1}{2} \int_{\partial B_R} \sum_{j=1}^N \sum_{j=1}^N x_i n_i (\nabla_j u)^2 \, d\sigma. \tag{29.10}$$

Combining (29.9) and (29.10), we have

$$-\int_{B_R} \Delta u \left(\sum_{i=1}^{N} x_i \nabla_i u\right) d\mathbf{x}$$
$$= \left(1-\frac{N}{2}\right)\int_{B_R} |\nabla u|^2\, d\mathbf{x} - R\int_{\partial B_R}\left|\frac{\partial u}{\partial \mathbf{n}}\right|^2 d\sigma + \frac{R}{2}\int_{\partial B_R} |\nabla u|^2\, d\sigma. \quad (29.11)$$

Thus, (29.8) and (29.11) imply

$$\int_{B_R} (-\Delta u + f(\mathbf{x},u)) \sum_{i=1}^{N} x_i \nabla_i u\, d\mathbf{x}$$
$$= R\left[\int_{\partial B_R}\left(F(\mathbf{x},u) + \frac{1}{2}|\nabla u|^2 - \left|\frac{\partial u}{\partial \mathbf{n}}\right|^2\right) d\sigma\right]$$
$$+ \left(1-\frac{N}{2}\right)\int_{B_R} |\nabla u|_2^2\, d\mathbf{x} - \int_{B_R}\left(N\, F(\mathbf{x},u) + \sum_{i=1}^{N} x_i \frac{\partial F(\mathbf{x},u)}{\partial x_i}\right) d\mathbf{x}. \quad (29.12)$$

Now we show that the boundary term on the left-hand side of (29.12) tends to zero for a subsequence $\{R_n\}$ with $R_n \to \infty$ as $n \to \infty$. First of all, note that $|\partial u/\partial \mathbf{n}| \le |\nabla u|$. We have, by the assumptions of the theorem,

$$\int_{\mathbb{R}^N} (|F(\mathbf{x},u)| + |\nabla u|^2)\, d\mathbf{x} = \int_0^{+\infty}\left(\int_{\partial B_R} (|F(\mathbf{x},u)| + |\nabla u|^2)\, d\sigma\right) dR < +\infty. \quad (29.13)$$

Consequently, there exists a subsequence $\{R_n\}$ with $R_n \to \infty$ as $n \to \infty$ such that

$$R_n \int_{\partial B_{R_n}} (|F(\mathbf{x},u)| + |\nabla u|^2)\, d\sigma \to 0 \qquad \text{as } n \to \infty. \quad (29.14)$$

Indeed, otherwise

$$\varliminf_{n\to\infty} R_n \int_{\partial B_R} (|F(\mathbf{x},u)| + |\nabla u|^2)\, d\sigma = C_0 > 0,$$

that is,

$$\lim_{n\to\infty} R_n \int_0^{+\infty}\left(\int_{\partial B_R} (|F(\mathbf{x},u)| + |\nabla u|^2)\, d\sigma\right) dR = +\infty,$$

which contradicts (29.13). Thus, because

$$\lim_{n\to\infty} \|\nabla u\|_2^2(B_{R_n}) = \|\nabla u\|_2^2,$$
$$\lim_{n\to\infty} \|F(\,.\,,u(\,.\,))\|_1(B_{R_n}) = \|F(\,.\,,u(\,.\,))\|_1,$$

(29.12) and (29.13) imply (29.7). □

Remark. According to the regularity theorems, the condition $u \in C^2(\mathbb{R}^N)$ may be weakened.

29.3 A FUNCTIONAL FACTOR OF NONEXISTENCE

THEOREM 29.5. *Let the conditions of Lemma 29.2 and Theorem 29.4 hold. Let for some $\alpha \in \mathbb{R}$ the integral relation*

$$\left(\frac{N-2}{2} - \alpha\right) \int_{\mathbb{R}^N} |\nabla u|_2^2 \, d\mathbf{x}$$
$$+ \int_{\mathbb{R}^N} \left(N\,F(\mathbf{x},u) + \sum_{i=1}^{N} x_i \frac{\partial F(\mathbf{x},u)}{\partial x_i} - \alpha\, u\, f(\mathbf{x},u) \right) d\mathbf{x} = 0 \qquad (29.15)$$

not be satisfied for $u \neq 0$ from the aforesaid class. Then the problem

$$-\Delta u + f(\mathbf{x},u) = 0 \qquad \text{on } \mathbb{R}^N,$$

has no (nontrivial) solution from the above class.

Proof. Multiply (29.7) by α and sum up with (29.4). The result follows. □

As an example, consider the problem

$$-\Delta u + \lambda_1 |u|^{p-2} u + \lambda_2 |u|^{q-2} u = 0 \qquad \text{on } \mathbb{R}^N, \qquad (29.16)$$

with some $1 < p \le q < +\infty$.

THEOREM 29.6. *Let $N > 2$ and $-\operatorname{sgn}(\lambda_1 (p - 2^*)) \neq \operatorname{sgn}(\lambda_2 (q - 2^*))$. Then problem (29.16) has no nontrivial classical solution from $\mathcal{E}_{p,q}$. Here*

$$\operatorname{sgn} z = \begin{cases} 1, & z > 0, \\ 0, & z = 0, \\ -1, & z < 0. \end{cases}$$

Proof. Letting $\alpha = \frac{N-2}{2}$ we lead identity (29.15) to the form

$$-\frac{N-2}{2}\left(\lambda_1 \|u\|_p^p + \lambda_2 \|u\|_q^q\right) + N\left(\frac{\lambda_1}{p}\|u\|_p^p + \frac{\lambda_2}{q}\|u\|_q^q\right) = 0.$$

Hence

$$-\lambda_1\left(\frac{1}{p}-\frac{1}{2^*}\right)\|u\|_p^p=\lambda_2\left(\frac{1}{q}-\frac{1}{2^*}\right)\|u\|_q^q,$$

and the statement of Theorem 29.6 becomes obvious. □

In accordance with the theorem we may divide the example into three cases. Problem (29.16) has no nontrivial solutions from $\mathcal{E}_{p,q}$ if

(i) $p<2^*$, $q<2^*$, and $\lambda_1\lambda_2\geq 0$,
(ii) $p>2^*$, $q>2^*$, and $\lambda_1\lambda_2\geq 0$,
(iii) $p<2^*$, $q>2^*$, and $\lambda_1\lambda_2\leq 0$.

Remark. Note that if $1<p<q<2^*$, $\lambda_1\geq 0$, and $\lambda_2\geq 0$, problem (29.16) has no nontrivial solution from $\mathcal{E}_{p,q}$ in accordance with Theorem 29.3. If $1<p\leq q<2^*$ and $\lambda_1>0$ and $\lambda_2<0$, problem (29.16) has a nontrivial solution from $\mathcal{E}_{p,q}$ in accordance with Theorem 9.10. If $p=2^*$, $\lambda_1>0$, and $\lambda_2=0$, problem (29.16) has a solution $u\in\mathcal{E}_{2^*}$ (see Section 22.6).

Remark. We emphasize that problem (29.16) has no solution in the indicated functional spaces. On the other hand, as we have seen in §27 the problem

$$-\Delta u-|u|^{p-2}u=0 \qquad \text{in } \mathbb{R}^N,$$

has nontrivial classical solutions for any $p\geq 2^*$ which, however, do not belong to $\mathcal{E}_p$.

Appendices

In this section we consider some results from classical functional analysis.

Appendix A. Spaces and Functionals

A.1 Main Definitions

Let X be a Banach space. Its norm will be denoted by $\|\,.\,\|$. Consider a set $\mathcal{P}$ of continuous linear mappings (functionals) $p : X \to \mathbb{R}$. The value of the functional p on an element $u \in X$ is denoted by

$$\langle p, u \rangle .$$

The linear space of functionals $p \in \mathcal{P}$ is called the dual space to X and is denoted by X^*.

Define a functional on $\mathcal{P}$,

$$\|p\|_* = \sup_{\|u\|=1} \langle p, u \rangle . \tag{A.1}$$

It is easy to see that it satisfies the axioms of a norm. The space X with the norm (A.1) is a Banach space.

Let $\{u_n\}$ be a sequence from X. We say that u_n converges to $u \in X$ strongly, $u_n \to u$ in X, if

$$\lim_{n\to\infty} \|u_n - u\| = 0.$$

We say that u_n converges to $u \in X$ weakly, $u_n \rightharpoonup u$ in X, if for any $v \in X^*$

$$\lim_{n\to\infty} \langle v, u_n \rangle = 0.$$

Let now $J : X \to \mathbb{R}$ be some, generally nonlinear, functional.

Definition. The functional $J : X \to \mathbb{R}$ is called continuously differentiable on X, denoted by $J \in C^1(X)$, if

(i) for any $u \in X$ there exists an element of X^* denoted by $J'(u)$ such that for all $v \in X$,

$$\lim_{t\to\infty} \frac{J(u+tv) - J(u) - t\langle J'(u), v\rangle}{t} = 0\,,$$

(ii) for any sequence $\{u_n\} \subset X$ strongly convergent in X,

$$\lim_{n\to\infty} \|J'(u_n) - J'(u)\|_* = 0.$$

The functional $J'(u)$ is called the derivative of J at the point u.

Continuous differentiability of a functional implies its continuousness.

A.2 The Functional T

The functional $T : \mathcal{E}_p \longrightarrow \mathbb{R}$,

$$T(u) = \|\nabla u\|_2^2,$$

was introduced in Section 1.3.

THEOREM A.1. *$T \in C^1(\mathcal{E}_p)$ and*

$$\langle T'(u), v\rangle = 2 \int\limits_{\mathbb{R}^N} (\nabla u, \nabla v)\, d\mathbf{x} \tag{A.2}$$

for all $v \in \mathcal{E}_p^*$.

Proof. First we verify condition (i) of the Definition. For this purpose we show that

(1) for any $u, v \in \mathcal{E}_p$

$$\lim_{t\to\infty} \frac{T(u+tv) - T(u) - 2\int_{\mathbb{R}^N} (\nabla u, \nabla v)\, d\mathbf{x}}{t} = 0,$$

(2) the functional $T'(u)$ defined by (A.2) belongs to $(\mathcal{E}_p)^*$.

Let us start with (1). We have

$$T(u+tv) - T(u) = \|\nabla(u+tv)\|_2^2 - \|\nabla u\|_2^2 = 2t \int\limits_{\mathbb{R}^N} (\nabla u, \nabla v)\, d\mathbf{x} + t^2 \|\nabla v\|_2^2.$$

Consequently, (1) holds.

Now we verify (2). The linearity of $T'(u)$ is obvious. Its continuousness follows from the observation that for any element $v \in \mathcal{E}_p$ and any sequence $\{u_n\} \in \mathcal{E}_p$ with

$$u_n \longrightarrow u \qquad \text{in } \mathcal{E}_p \qquad \text{as } n \longrightarrow \infty,$$

the Hölder inequality implies

$$|\langle T'(v), u_n - u\rangle| = 2 \left| \int\limits_{\mathbb{R}^N} (\nabla v, \nabla(u_n - u))\, d\mathbf{x} \right| \le 2\|\nabla v\|_2 \, \|\nabla(u_n - u)\|_2 \longrightarrow 0.$$

Now we check (ii) of the Definition. For any element $v \in \mathcal{E}_p$ and any sequence $\{u_n\} \in \mathcal{E}_p$ with

$$u_n \longrightarrow u \qquad \text{in } \mathcal{E}_p \qquad \text{as } n \longrightarrow \infty,$$

the Hölder inequality yields

$$|\langle T'(u) - T'(u_n), v\rangle| = 2 \left| \int\limits_{\mathbb{R}^N} (\nabla(u_n - u), \nabla v)\, d\mathbf{x} \right| \le 2\, \|\nabla(u_n - u)\|_2 \|\nabla v\|_2,$$

which implies

$$\sup_{\|\nabla v\|_2 + \|v\|_p = 1} \langle T'(u) - T'(u_n), v\rangle \le 2\, \|\nabla(u_n - u)\|_2 \longrightarrow 0.$$

Therefore, the theorem is proved. □

A.3 THE FUNCTIONAL V

The functional $V : \mathcal{E}_{p,q} \to \mathbb{R}$,

$$V(u) = \int_{\mathbb{R}^N} F(\mathbf{x}, u)\, d\mathbf{x},$$

where

$$F(\mathbf{x}, t) = \int_0^t f(\mathbf{x}, s)\, ds$$

was introduced in Section 1.3 for the Carathéodory function $f \in K_{p,q}$.

THEOREM A.2. *$V \in C^1(\mathcal{E}_{p,q})$ and*

$$\langle V'(u), v\rangle = \int_{\mathbb{R}^N} f(\mathbf{x}, u) v\, d\mathbf{x} \tag{A.3}$$

for all $v \in \mathcal{E}_p^*$.

Proof. The proof is in the spirit of the proof of Theorem A.1 but is more cumbrous. As an illustration we show that condition (ii) is valid.

Let $\{u_n\} \in \mathcal{E}_{p,q}$ be a sequence strongly convergent to u in $\mathcal{E}_{p,q}$. Then from the Hölder inequality and $f \in K_{p,q}$, we get

$$\begin{aligned}
|\langle V'(u_n) - V'(u), v\rangle| &= \left| \int_{\mathbb{R}^N} (f(\mathbf{x}, u_n) - f(\mathbf{x}, u)) v\, d\mathbf{x} \right| \\
&\leq \left| \int_{B_R} (f(\mathbf{x}, u_n) - f(\mathbf{x}, u)) v\, d\mathbf{x} \right| \\
+ \Bigg| \int_{\mathbb{R}^N \setminus B_R} &(\mathbf{C}(|u|^{p-1} + |u|^{q-1} + |u_n|^{p-1} + |u_n|^{q-1}) + f_1(\mathbf{x}) + f_2(\mathbf{x})) v\, d\mathbf{x} \Bigg| \\
&\leq \left| \int_{B_R} (f(\mathbf{x}, u_n) - f(\mathbf{x}, u)) v\, d\mathbf{x} \right| \\
&+ \mathbf{C}\, (\|u\|_p^{p-1} + \|u_n\|_p^{p-1} + \|f_1\|_{p/(p-1)})(\mathbb{R}^N \setminus B_R) \|v\|_p \\
&\qquad + (\|u\|_q^{q-1} + \|u_n\|_q^{q-1} + \|f_2\|_{q/(q-1)})(\mathbb{R}^N \setminus B_R) \|v\|_q,
\end{aligned}$$

where the constants $\mathbf{C}$ do not depend on n and R and the functions f_1 and f_2 are defined by $f \in K_{p,q}$.

Because $\{u_n\}$ converges strongly to u in $\mathcal{E}_{p,q}$, for any $\varepsilon > 0$ there is n_1 such that for all $n \geq n_1$ there exists an R such that

$$\mathbf{C}\,(\|u\|_p^{p-1} + \|u_n\|_p^{p-1} + \|f_1\|_{p/(p-1)})(\mathbb{R}^N \setminus B_R)$$

and

$$\mathbf{C}\,(\|u\|_q^{q-1} + \|u_n\|_q^{q-1} + \|f_1\|_{q/(q-1)})(\mathbb{R}^N \setminus B_R)$$

are less than ε. On the other hand, $f \in \mathcal{E}_q(B_R)$ for $1 \leq p \leq q$ because

$$\|f\|_p^p(B_R) \leq \|f\|_q^q(B_R)\|1\|^{1-p/q}(B_R),$$
$$|t|^p \leq |t|^q + 1, \qquad 1 \in \mathcal{E}_q(B_R).$$

This means that we can apply the Vainberg-Krasnoselskii theorem and find $n_0 \geq n_1$ such that for all $n \geq n_0$,

$$\left|\int_{B_R} (f(\mathbf{x}, u_n) - f(\mathbf{x}, u))v\, d\mathbf{x}\right|$$
$$\leq \left|\int_{B_R} |f(\mathbf{x}, u_n) - f(\mathbf{x}, u)|^{\frac{q-1}{q}}\, d\mathbf{x}\right|^{\frac{q}{q-1}} \|v\|_q(B_R) \leq \varepsilon\|v\|_q(B_R).$$

Thus, given ε, we can choose R and then n_0 such that for all $n \geq n_0$,

$$\|V'(u_n) - V'(u)\|_* \leq \sup_{\|v\|_p + \|v\|_q = 1} \langle V'(u_n) - V(u), v\rangle \leq \varepsilon\,\mathbf{C}.$$

Letting $\varepsilon \to \infty$ we get (ii) for V. □

Appendix B. The Strauss Lemma

In this section the Strauss lemma is proved.

THE STRAUSS LEMMA. *Let $P, Q : \mathbb{R} \to \mathbb{R}$ be continuous functions with $Q(s) \neq 0$ for $s \neq 0$ such that*

$$\lim_{|s|\to 0} \left|\frac{P(s)}{Q(s)}\right| = 0, \qquad \lim_{|s|\to\infty} \left|\frac{P(s)}{Q(s)}\right| = 0. \tag{B.1}$$

Then for any sequence $\{u_n\}$ of functions measurable on $\mathbb{R}^N$ satisfying

(1) $u_n \to u$ *almost everywhere in* $\mathbb{R}^N$ *as* $n \to \infty$,
(2) $\|Q(u_n)\|_1 \leq \mathbf{C}$
(3) $u_n \to 0$ *as* $|\mathbf{x}| \to +\infty$ *uniformly with respect to* n,
there holds

$$P(u_n) \to P(u) \qquad \text{in } L_1(\mathbb{R}^N) \qquad \text{as } n \to \infty.$$

This lemma is based on the following result generalizing the Vainberg-Krasnoselskii theorem.

LEMMA B.1. *Let $P, Q : B_R \to \mathbb{R}$ be continuous functions with $Q(s) \neq 0$ for $s \neq 0$ such that*

$$\lim_{|s|\to\infty} \left| \frac{P(s)}{Q(s)} \right| = 0.$$

Then, for any sequence $\{u_n\}$ of functions measurable on B_R satisfying

(1) $u_n \to u$ *almost everywhere in* B_R *as* $n \to \infty$,

(2) $\|Q(u_n)\|_1 \leq \mathbf{C}$ *uniformly with respect to* n,

the following holds

$$P(u_n) \to P(u) \qquad \text{in } L_1(B_R) \qquad \text{as } n \to \infty.$$

Proof of the Strauss Lemma. By virtue of the continuousness of functions P and Q and assumption *(1)* we have

$$P(u_n) \to P(u) \qquad \text{a.e. in } \mathbb{R}^N. \tag{B.2}$$

From (B.1) the function $|P/Q|$ is continuous and attains its maximum, that is, there is a constant $C > 0$ such that for all $t \in \mathbb{R}$,

$$|P(t)| \leq C\,|Q(t)|.$$

Then, by assumption *(2)*,

$$\|P(u_n)\|_1 \leq C\,\|Q(u_n)\|_1 \leq \mathbf{C}$$

with $\mathbf{C}$ not depending on n. From the Lebesgue-Fatou theorem, this estimate and (B.2) imply

$$P(u) \in L_1(\mathbb{R}^N).$$

In accordance with *(1)* and *(3)* for any $\delta > 0$ there is a radius R_δ and a number n_δ such that for all $n > n_\delta$,

$$|u_n(\mathbf{x})| \leq \delta, \qquad |u(\mathbf{x})| \leq \delta \qquad \text{for } |\mathbf{x}| \geq R_\delta. \tag{B.3}$$

From (B.1) and (B.3), for any ε there is a radius R_ε and a number n_ε such that for all $n > n_\varepsilon$

$$\|P(u_n)\|_1\,(\mathbb{R}^N \setminus B_{R_\varepsilon}) \leq \varepsilon \|Q(u_n)\|_1\,(\mathbb{R}^N \setminus B_{R_\varepsilon}) \leq \mathbf{C}\,\varepsilon$$

and, analogously,

$$\|P(u)\|_1\,(\mathbb{R}^N \setminus B_{R_\varepsilon}) \leq \mathbf{C}\,\varepsilon$$

with $\mathbf{C}$ not depending on ε and n.

From Lemma B.1 we may find a number $n_\varepsilon^0 \geq n_\varepsilon$ such that

$$\|P(u_n) - P(u)\|_1\,(B_R) \leq \varepsilon$$

for $n > n_\varepsilon^0$. In other words, for any $\varepsilon > 0$ one can find a radius R_ε and a number n_ε^0 such that

$$\begin{aligned} &\|P(u_n) - P(u)\|_1 \\ &\leq \|P(u_n) - P(u)\|_1\,(B_R) + \|P(u_n)\|_1\,(\mathbb{R}^N \setminus B_{R_\varepsilon}) + \|P(u)\|_1\,(\mathbb{R}^N \setminus B_{R_\varepsilon}) \leq \mathbf{C}\,\varepsilon \end{aligned}$$

with $\mathbf{C}$ not depending on ε and n.

Letting $\varepsilon \to 0$ we get the result. □

Appendix C. Invariant Spaces

C.1 Radial Case

In this section we dwell upon results connected with invariance properties of functionals. We recall that throughout the book we assumed that f is a Carathéodory function.

Let $f : \mathbb{R}^N \times \mathbb{R} \to \mathbb{R}$ be a function of the class $K_{p,q}$. Then from Section 1.4 the functional $\Phi : \mathcal{E}_{p,q} \to \mathbb{R}$ of the form

$$\Phi(u) = \frac{1}{2}\|\nabla u\|_2^2 + \int_R F(\mathbf{x}, u)\, d\mathbf{x} \tag{C.1}$$

is well-defined, where

$$F(\mathbf{x}, t) = \int_0^t f(\mathbf{x}, s)\, ds.$$

Suppose that f depends on $\mathbf{x}$ only through $|\mathbf{x}|$, that is, f is radial. Define the space $\mathcal{E}_{p,q}^{\mathrm{rad}}$ as the subspace of $\mathcal{E}_{p,q}$ consisting of radial functions. Define without changing notation the functional $\Phi : \mathcal{E}_{p,q}^{\mathrm{rad}} \to \mathbb{R}$ by the same relation (C.1).

Theorem C.1. *Let $f \in K_{p,q}$ be radial. Let u be a radial critical point of the functional Φ considered on $\mathcal{E}_{p,q}^{\mathrm{rad}}$. Then u is also a critical point of this functional considered on $\mathcal{E}_{p,q}$.*

Proof. Because of density reasons it is sufficient to prove that if the identity

$$\langle \Phi'(u), v\rangle = 0$$

is valid for all radial $v \in \mathcal{D}(\mathbb{R}^N)$, then it remains true for arbitrary $v \in \mathcal{D}(\mathbb{R}^N)$.

Given $v \in \mathcal{D}(\mathbb{R}^N)$, define functions $v_1, v_2 \in \mathcal{D}(\mathbb{R}^N)$ as

$$v_1(r) = S_N^{-1} \int_{\partial B_r} v(\mathbf{x})\, d\sigma, \qquad v_2 = v - v_1$$

where $d\sigma$ is an area element, S_N is the area of the unit sphere in $\mathbb{R}^N$, and $r = |\mathbf{x}|$. Obviously, $\langle \Phi'(u), v_1\rangle = 0$ because v_1 is radial and u is a critical point of Φ considered on the radial space.

On the other hand,

$$\begin{aligned}
&\langle \Phi'(u), v_2\rangle \\
&\quad = \int_0^\infty r^{N-1} dr \int_{\partial B_R} d\sigma\, (\nabla u(r), \nabla(v(\mathbf{x}) - v_1(r)) + f(r, u(r))(v(\mathbf{x}) - v_1(r))) \\
&= S_N \int_0^\infty r^{N-1} dr\, ((\nabla_r u(r), \nabla_r(v_1(r) - v_1(r))) + f(r, u(r))(v_1(r) - v_1(r))) = 0.
\end{aligned}$$

The theorem is proved. □

Let now G be a subgroup of the orthogonal group acting as described in §16. In this section G-symmetric functions were also defined. A result analogous to Theorem C.1 is true for such G-symmetric functions.

THEOREM C.2. *Let $f \in K_{p,q}$ be G-symmetric. Let u be a G-symmetric critical point of the functional Φ considered on $\mathcal{E}^G_{p,q}$. Then u is also a critical point of this functional considered on $\mathcal{E}_{p,q}$.*

Proof. It is analogous to the preceding one. □

C.2 ANTI-BLOCK-RADIAL SYMMETRY

Let $N = 4$ or $N > 5$. Let $f : \mathbb{R} \to \mathbb{R}$ be a continuous odd function. Define the functionals Φ by (C.1) with radial f.

Here we use the notation and definitions at the beginning of Section 17.2.

THEOREM C.3. *Let $u \in \mathcal{E}^{(s_1,\dots,[s_i,s_j],\dots,s_m)}_{p,q}$ be a critical point of the functional Φ considered on $\mathcal{E}^{(s_1,\dots,[s_i,s_j],\dots,s_m)}_{p,q}$. Then it is a critical point of this functional considered on $\mathcal{E}^{(s_1,\dots,s_m)}_{p,q}$.*

Note that any function $u \in \mathcal{E}^{(s_1,\dots,s_m)}_{p,q}$ is G-symmetric with an appropriate G and hence Theorems C.2 and C.3 imply

THEOREM C.4. *Let $u \in \mathcal{E}^{(s_1,\dots,[s_i,s_j],\dots,s_m)}_{p,q}$ be a critical point of the functional Φ considered on $\mathcal{E}^{(s_1,\dots,[s_i,s_j],\dots,s_m)}_{p,q}$. Then it is a critical point of this functional considered on $\mathcal{E}_{p,q}$.*

Proof of Theorem C.3. Assume for simplicity that $m = 2$, $i = 1$, and $j = 2$. Density reasons allow us to restrict ourselves to functions from $\mathcal{D}(\mathbb{R}^N)$.

Given $v \in \mathcal{D}(\mathbb{R}^N)$, define two functions $v_1, v_2 \in \mathcal{D}(\mathbb{R}^N)$ as

$$v_1(s_1,s_2) = \frac{1}{2}(v(s_1,s_2) - v(s_2,s_1)),$$
$$v_2(s_1,s_2) = \frac{1}{2}(v(s_1,s_2) + v(s_2,s_1)).$$

Obviously,

$$v = v_1 + v_2, \qquad v_1(s_1,s_2) = -v_1(s_2,s_1), \qquad v_2(s_1,s_2) = v_2(s_2,s_1). \tag{C.2}$$

Let $u \in \mathcal{E}^{([s_1,s_2])}_{p,q}$ be a critical point of the functional Φ considered on $\mathcal{E}^{([s_1,s_2])}_{p,q}$. Then

$$\langle \Phi'(u), v_1 \rangle = 0$$

follows from $v_1 \in \mathcal{E}_{p,q}^{([s_1,s_2])}$. On the other hand, by (C.2),

$$\langle \Phi'(u), v_2\rangle = S_{N/2}^2 \int_0^\infty \int_0^\infty ((\nabla u, \nabla v_2) + f(u)\, v_2)\, s_1^{N/2-1} ds_1\, s_2^{N/2-1} ds_2$$

$$= \frac{1}{2} S_{N/2}^2 \int_0^\infty \int_0^\infty ((\nabla (u(s_1,s_2) - u(s_2,u_1)), \nabla v_2(s_1,s_2))$$

$$+ (f(u(s_1,s_2)) - f(u(s_2,s_1)))\, v_2(s_1,s_2))\, s_1^{N/2-1} ds_1\, s_2^{N/2-1} ds_2$$

$$= \frac{1}{2} S_{N/2}^2 \Big(\int_0^\infty \int_0^\infty ((\nabla u(s_1,s_2), \nabla v_2(s_1,s_2))$$

$$+ f(u(s_1,s_2))\, v_2(s_1,s_2))\, s_1^{N/2-1} ds_1\, s_2^{N/2-1} ds_2$$

$$- \int_0^\infty \int_0^\infty ((\nabla u(s_2,s_1), \nabla v_2(s_2,s_1))$$

$$+ f(u(s_2,s_1))\, v_2(s_2,s_1))\, s_1^{N/2-1} ds_1\, s_2^{N/2-1} ds_2 \Big) = 0.$$

Thus,

$$\langle \Phi'(u), v\rangle = \langle \Phi'(u), v_1\rangle + \langle \Phi'(u), v_2\rangle = 0$$

for any $v \in \mathcal{E}_{p,q}^{(s_1,s_2)}$, which proves the theorem. □

Appendix D. The Schwarz Rearrangement

An extremum of the symmetric functionals in the corresponding Banach spaces is as a rule attained on the functions also possessing these symmetries. Here we consider a procedure of "symmetrization" which is helpful in finding such functions for elliptic equations with the Laplace operator.

D.1 Definitions and Main Properties

Let $u : \mathbb{R}^N \to \mathbb{R}$ be a measurable function, and for any real $c > 0$,

$$\text{mes}\,\{\mathbf{x} \in \mathbb{R}^N,\ |u(\mathbf{x})| \geq c\}$$

be finite.

Definition. The function u^* is a rearranged one for u by Schwarz if u^* is radial, nonnegative, and for any $c > 0$,

$$\text{mes}\,\{\mathbf{x} \in \mathbb{R}^N,\ u^*(\mathbf{x}) \geq c\} = \text{mes}\,\{\mathbf{x} \in \mathbb{R}^N,\ |u(\mathbf{x})| \geq c\}. \tag{D.1}$$

Functions rearranged by Schwarz have some interesting properties.

Suppose that $F : \mathbb{R} \to \mathbb{R}$ is a continuous even nonnegative function, $F \in K_{p+1,q+1}$ with $1 < p, q < +\infty$.

THEOREM D.1. *Let $u \in \mathcal{E}_{p,q}$. Then $u^* \in \mathcal{E}_{p,q}$ and*

$$\int_{\mathbb{R}^N} F(u^*)\,d\mathbf{x} = \int_{\mathbb{R}^N} F(u)\,d\mathbf{x}.$$

Let now $a : \mathbb{R}^+ \to \mathbb{R}$ be a nonnegative continuous nondecreasing function.

THEOREM D.2. *Let $u \in \mathcal{E}_{p(a)}$. Then $u^* \in \mathcal{E}_{p(a)}$, and*

$$\int_{\mathbb{R}^N} a(|\mathbf{x}|)|u^*|^p\,d\mathbf{x} \leq \int_{\mathbb{R}^N} a(|\mathbf{x}|)|u|^p\,d\mathbf{x}.$$

THEOREM D.3. *Let $u \in \mathcal{E}_p$. Then $u^* \in \mathcal{E}_p$, and*

$$\|\nabla u^*\|_2^2 \leq \|\nabla u\|_2^2.$$

It is easy to see that Theorems D.1 and D.3 imply

THEOREM D.4. *Let $F : \mathbb{R} \to \mathbb{R}$ be a continuous even nonnegative function, $F \in K_{p+1,q+1}$ with $1 < p,\ q < +\infty$, and*

$$J_1(u) = \|\nabla u\|_2^2,$$

$$J_2(u) = \int_{\mathbb{R}^N} F(u)\,d\mathbf{x}.$$

Let the set

$$\{u \in \mathcal{E}_{p,q},\ J_2(u) = C_0\}$$

be not empty for a constant C_0.

Then

$$\inf_{\substack{u\in\mathcal{E}_{p,q}\\ J_2(u)=C_0}} J_1(u) = \inf_{\substack{u\in\mathcal{E}^{\mathrm{rad}}_{p,q}\\ J_2(u)=C_0}} J_1(u).$$

If the problem

$$\{u \in \mathcal{E}_{p,q},\ \min J_1(u),\ J_2(u) = C_0\}$$

has a solution u, then $u = u^*$.

Theorems D.2 and D.3 give rise to an analog of Theorem D.4.

THEOREM D.5. *Let* $a : \mathbb{R}^+ \to \mathbb{R}$ *be a nonnegative continuous nondecreasing function, and*

$$J_1(u) = \|\nabla u\|_2^2,$$

$$J_2(u) = \int\limits_{\mathbb{R}^N} a\,|u|^p\, d\mathbf{x}.$$

Let the set

$$\{u \in \mathcal{E}_{p,q},\ J_2(u) = C_0\}$$

be not empty for a constant C_0.

Then

$$\inf_{\substack{u\in\mathcal{E}_{p,q}\\ J_2(u)=C_0}} J_1(u) = \inf_{\substack{u\in\mathcal{E}^{\mathrm{rad}}_{p,q}\\ J_2(u)=C_0}} J_1(u)\,.$$

If the problem

$$\{u \in \mathcal{E}_{p,q},\ \min J_1(u),\ J_2(u) = C_0\}$$

has a solution u, *then* $u = u^*$.

D.2. REARRANGEMENT FOR POLYNOMIALS IN $\mathbb{R}$. EXPLICIT CONSTRUCTIONS

We present the proof of Theorems D.1–D.3 only in the simple case where u has a special form and $N = 1$. The general case requires an appropriate approximation procedure and lies beyond the scope of our book. For a more detailed treatment see Kawohl [1].

Let u be a nonnegative polynomial of degree n defined on some interval $[a, b]$, and

$$u(a) = u(b) = 0.$$

It is obvious that if $\max_x u > C > 0$, the equation

$$u(x) = C$$

in x has no less than two but no more than n solutions. Analogously, the equation

$$u'(x) = 0$$

in x has no more than $n - 1$ solutions.

This means that the interval $(0, \max_x u(x))$ can be divided into a finite number of intervals (C_i, C_{i+1}), where C_i, for $i \geq 2$, are real numbers for which roots of

$$u'(x) = 0, \qquad u(x) = C_i$$

exist. Enumerate them in increasing order, letting $C_1 = 0$ and $C_{\max} = \max_x u(x)$.

Let $C \in (C_i, C_{i+1})$. Then enumerate roots of $u(x) = C$ in increasing order, too. Denote their quantity as $\mathcal{N}(C)$.

Now we are ready to build a rearrangement. Given $u : x \mapsto u$, with the domain $[a, b]$ and range $[0, \max_x u(x)]$, we may define a set of $\mathcal{N}(C)$ functions $x_j : u \mapsto x$, inverse to u on each interval (C_i, C_{i+1}). Given a set of x_j, we may define a function

$$x^* : (C_i, C_{i+1}) \to \left(0, \frac{b-a}{2}\right)$$

as

$$x^*(u) = \frac{1}{2}\left|\sum_{j=1}^{\mathcal{N}(C)} (-1)^j x_j(u)\right|.$$

Define x^* at points C_i by continuousness. It is easy to see that

$$x^*(C) = \frac{1}{2}\text{mes}\,\{x \in (a, b),\ u(x) \geq C\}$$

and thus this function is nonincreasing in C.

This implies that, given $x^* : u \mapsto x$, the function $u^* : x \mapsto u$ can be defined as follows: u^* is inverse to x^* if $x > 0$ and is inverse to $-x^*$ if $x < 0$. Let $u^*(0) = \max_x u(x)$.

Thus, the function $u^* : (a, b) \to [0, \max_x u(x)] \subset \mathbb{R}$ is rearranged by the Schwarz method for u.

Now we are ready to verify Theorems D.1–D.3 in our simple case.

PROOF OF THEOREMS D.1 AND D.2

Theorem D.1 follows from the definition of the Lebesgue integral and relation (D.1):

$$\int_a^b F(u(\mathbf{x}))\,d\mathbf{x} = \int_0^\infty \text{mes}\,\{\mathbf{x} \in \mathbb{R},\ F(u(\mathbf{x})) \geq c\}\,dc$$

$$= \int_0^\infty \text{mes}\,\{\mathbf{x} \in \mathbb{R},\ F(u^*(\mathbf{x})) \geq c\}\,dc = \int_{\frac{a-b}{2}}^{\frac{b-a}{2}} F(u^*(\mathbf{x}))\,d\mathbf{x}.$$

Theorem D.2 may be proved analogously. □

PROOF OF THEOREM D.3, SIMPLIFIED VERSION

In the proof of our simplified case $N = 1$ we shall follow Polya and Szego [1]. The strict modern approach may be found in Brascamp, Lieb, and Luttinger [1].

Preliminarily, we prove an auxiliary result.

LEMMA D.6. *For any nonnegative a_i and θ_i the inequality*

$$\sum_i \theta_i^2 a_i \geq \frac{\left(\sum_i \theta_i a_i\right)^2}{\sum_i a_i} \tag{D.2}$$

holds.

Proof. Inequality (D.2) follows from the following observation:

$$\sum_i \theta_i a_i = \sum_i \theta_i \sqrt{a_i}\sqrt{a_i} \leq \left(\sum_i \theta_i^2 a_i\right)^{1/2} \left(\sum_i a_i\right)^{1/2}. \qquad \square$$

Now we get

$$\int_a^b \left|\frac{\partial u(x)}{\partial x}\right|^2 dx \overset{\text{by change of variables}}{=} \sum_i \int_{C_i}^{C_{i+1}} \sum_{j=1}^{\mathcal{N}(C)} \left|\frac{\partial x_j(u)}{\partial u}\right|^{-2} \left|\frac{\partial x_j(u)}{\partial u}\right| du$$

$$\overset{(D.2)}{\geq} \sum_i \int_{C_i}^{C_{i+1}} \frac{\left(\sum_{j=1}^{\mathcal{N}(C)} 1\right)^2}{\sum_{j=1}^{\mathcal{N}(C)} \left|\frac{\partial x_j(u)}{\partial u}\right|}$$

$$\overset{\mathcal{N}(C)\geq 2}{\geq} \sum_i \int_{C_i}^{C_{i+1}} \frac{4}{\sum_{j=1}^{\mathcal{N}(C)} \left|\frac{\partial x_j(u)}{\partial u}\right|}$$

$$= \sum_i \int_{C_i}^{C_{i+1}} \frac{du}{\left|\frac{\partial x^*(u)}{\partial u}\right|} = \int_{\frac{a-b}{2}}^{\frac{b-a}{2}} \left|\frac{\partial u^*(x)}{\partial x}\right|^2 dx.$$

Thus, Theorem D.3 is proved for nonnegative polynomials in the case $N = 1$. $\square$

Appendix E. The Mountain Pass Method

In this section, proofs of the theorems on the mountain pass method are given. In the beginning of the section a so-called deformation lemma is proved, which is the crucial tool in this and many other approaches beyond this book.

We follow Ambrosetti and Rabinowitz [1], Clark [1], Palais [1–3], Palais and Smale [1]. A review of modern results on this subject may be found in the beautiful survey of Rabinowitz [1].

E.1 Deformation Lemma. Formulation

Let X be a real Banach space and $J : X \to \mathbb{R}$ be a functional of the class C^1.

Introduce the following notation:

$$K_{a,\varepsilon} = \{u \in X,\ |J(u) - a| \leq \varepsilon\},$$
$$A_a = \{u \in X,\ J(u) \leq a\}.$$

THEOREM E.1. *(Deformation lemma) Let there exist constants a, $C > 0$, $\varepsilon > 0$ such that for all $u \in K_{a,\varepsilon}$ the inequality*

$$\|J'(u)\|_* \geq C \tag{E.1}$$

holds.

Then for any δ with $\varepsilon > \delta > 0$ there is a continuous mapping $\eta : [0,1] \times X \to X$, which is a mapping of X onto itself for any $t \in [0,1]$, such that

$$\begin{array}{lll} (A) & \eta(t,u) = u & \text{for all } u \notin K_{a,\varepsilon}, \\ (B) & \eta(0,u) = u & \text{for all } u \in X, \\ (C) & \eta(1, A_{a+\delta}) \subset A_{a-\delta}. & \end{array}$$

E.2 Proof of a Simplified Deformation Lemma

First of all, in order to underline the main idea we give the proof of a simplified version of Theorem E.1. Suppose that $X = X^*$ and J belongs to the class $C^2(X)$.

Let us start. Define a Lipschitz function $g : \mathbb{R} \to \mathbb{R}^+$ by

$$g(t) = \begin{cases} 0 & \text{if } t \notin [a-\varepsilon, a+\varepsilon], \\ 1 & \text{if } t \in [a-\delta, a+\delta], \\ \text{linear} & \text{otherwise,} \end{cases} \tag{E.2}$$

and consider the equation

$$\begin{aligned} &\frac{d\bar{\eta}}{dt} = -2\delta g(J(\bar{\eta}))\frac{J'(\bar{\eta})}{\|J'(\bar{\eta})\|^2} \qquad \text{in } \mathbb{R}^+, \\ &\bar{\eta}(t=0) = u, \end{aligned} \tag{E.3}$$

where the right-hand part of the equation (E.3) is supposed zero at points where $J'(\bar{\eta}) = 0$.

It is known from functional analysis that if the right-hand side of the first relation of (E.3) is locally Lipschitz and uniformly bounded for all $\bar{\eta} \in X$, then (E.3) has a unique solution $\bar{\eta} \in X$, and the functional $\eta : [0,1] \times X \to X$ defined as

$$\eta(t,u) = \bar{\eta}(t) \quad \text{with } \bar{\eta}(t=0) = u \tag{E.4}$$

is continuous in t and u (e.g., Berger [2]).

The right-hand side is locally Lipschitz by definition. Verify that it is uniformly bounded. Indeed, from (E.1) and $X^* = X$, we have

$$\left\| 2\delta g(J(\eta(t,u))) \frac{J'(\eta(t,u))}{\|J'(\eta(t,u))\|^2} \right\| \leq 2\delta \|J'(\eta(t,u))\|^{-1} \leq 2\delta C^{-1} \qquad \text{(E.5)}$$

for all $u \in X$.

Let us show that η satisfies assumptions (A), (B), (C). Assumption (A) follows directly from the definition. Assumption (B) is true because the right-hand side of (E.3) is equal to zero for all $u \notin K_{a,\varepsilon}$ thanks to (E.2).

As for assumption (C), we have for all t and u,

$$J(\eta(t,u)) - J(\eta(0,u)) = \int_0^1 \frac{dJ(\eta(t,u))}{dt}\, dt\,. \qquad \text{(E.6)}$$

By differentiation, we have by force of equation (E.3)

$$\frac{dJ(\eta)}{dt} = \left\langle J'(\eta), \frac{d\eta}{dt} \right\rangle = -2\delta\, g(J(\eta)) \left\langle J'(\eta), \frac{J'(\eta)}{\|J'(\eta)\|^2} \right\rangle = -2\delta\, g(J(\eta)). \qquad \text{(E.7)}$$

Relations (E.6) and (E.7) imply that for $u \in K_{a,\delta}$,

$$J(\eta(1,u)) = J(\eta(0,u)) - \int_0^1 2\delta\, dt \leq a + \delta - 2\delta = a - \delta. \qquad \text{(E.8)}$$

Analogously, for all $u \in A_{a-\delta}$,

$$J(\eta(1,u)) = J(\eta(0,u)) - 2\delta \int_0^1 g(J(\eta(t,u(t))))\, dt \leq J(\eta(0,u)) \leq a - \delta. \qquad \text{(E.9)}$$

Relations (E.8) and (E.9) imply (C). □

E.3 A PSEUDOGRADIENT FIELD

In the general case, the complication in the proof is connected with the fact that $J'(u)$ does not belong to X whereas the right-hand side of equation (E.3) must lie in X. A lesser smoothness of J also does not simplify the matter. In order to overcome these difficulties the notion of a pseudogradient field is introduced, which is due to Palais [3].

DEFINITION. Let X be a Banach space and a functional $J : X \to \mathbb{R}$ of the class C^1 be given. A functional $W : X \to X$ is called a pseudogradient field if the following holds:

(α) $\|W(u)\| \leq 2\,\|J'(u)\|_*$ for all $u \in X$,

(β) $\langle J'(u), W(u)\rangle \geq \|J'(u)\|_*^2$ for all $u \in X$,

(γ) the functional W is locally Lipschitz.

THEOREM E.2. *Let $J \in C^1(X)$. Then there is a pseudogradient field of J.*

Proof. Let u_0 be an arbitrary point in X. In accordance with the definition of a norm there exists an element $v \in X$ such that

$$\langle J'(u_0), v\rangle \geq \frac{3}{4}\|J'(u_0)\|_*, \qquad \|v\| = 1. \tag{E.10}$$

Letting

$$W_0(u_0) = \frac{3}{2}\|J'(u_0)\|_* \cdot v \tag{E.11}$$

we get, substituting (E.10) into (E.11),

$$\langle J'(u_0), W_0(u_0)\rangle \geq \frac{9}{8}\|J'(u_0)\|_*^2. \tag{E.12}$$

Because the values of $J'(u)$ depend on u continuously in X^*, relations (E.11), (E.12) mean that in some neighborhood U_ε of u_0

$$\begin{gathered} W_0(u_0) \leq 2\,\|J'(u_0)\|_*, \\ \langle J'(u_0), W_0(u_0)\rangle \geq \|J'(u_0)\|_*^2 \end{gathered} \tag{E.13}$$

provided we let

$$W_0(u) = W_0(u_0) \qquad \text{for } u \in U_\varepsilon. \tag{E.14}$$

Thus, we may cover X by open domains U_ε^i on which constant functionals W_i satisfying the analogs of (E.13) and (E.14) for appropriate u_i are given. General topology says (see Engelking [1]) that in any open covering $\{U_\varepsilon^i\}$ of a Banach space one can inscribe an open locally finite covering $\{V_\varepsilon^i\}$, i.e., such a covering that for any $u \in X$ there exists a finite set of indexes i with $u \in V_\varepsilon^i$.

Now, our task is to sew together the covering $\{V_\varepsilon^i\}$ in such a way that we get a locally Lipschitz functional. For this purpose we consider the following functionals, $g_i : V_\varepsilon^i \to \mathbb{R}$,

$$g_i(u) = \inf_{v \in \partial V_\varepsilon^i} \|u - v\|$$

and let

$$g_i(u) = 0 \qquad \text{outside } V_\varepsilon^i.$$

Functionals g_i are locally Lipschitz. Indeed, from the axioms of norms,

$$\begin{aligned} \|g_i(u) - g_i(v)\| &= \left| \inf_{u' \in \partial V_\varepsilon^i} \|u - u'\| - \inf_{v' \in \partial V_\varepsilon^i} \|v - v'\| \right| \\ &\leq \inf_{\substack{u' \in \partial V_\varepsilon^i \\ v' \in \partial V_\varepsilon^i}} \|u - u' - v + v'\| \leq \|u - v\|. \end{aligned}$$

Now we define the functional $W : X \to X$ by

$$W(u) = \frac{\sum_i g_i(u)\, W_i(u)}{\sum_i g_i(u)}. \tag{E.15}$$

and show that it is a pseudogradient field. □

Verifying (γ). By local finiteness of the covering, the numerator and denominator are Lipschitz as finite sums of Lipschitz terms. Moreover the denominator can not equal zero because all V_ε^i are open and every point belongs to at least one V_ε^i. This means that the fracture is also locally Lipschitz. □

Verifying (α). From the norm axioms and (E.13) we have

$$\|W(u)\| \le \frac{\sum_i g_i(u)\, \|W_i(u)\|}{\sum_i g_i(u)} \le 2\frac{\sum_i g_i(u)\, \|J'(u)\|_*}{\sum_i g_i(u)} = 2\|J'(u)\|_*$$

for all $u \in X$. □

Verifying (β). We have analogously, by (E.13),

$$\langle J'(u), W(u)\rangle = \frac{\sum_i g_i(u)\, \langle J'(u), W_i(u)\rangle}{\sum_i g_i(u)} \ge \|J'(u)\|_*^2.$$ □

E.4 The Full Proof of the Deformation Lemma

Analogously to (E.3) we consider the equation

$$\begin{aligned} \frac{d\bar\eta}{dt} &= -2\delta g(J(\bar\eta))\frac{W(\bar\eta)}{\|J'(\bar\eta)\|^2} \qquad \text{in } \mathbb{R}^+, \\ \bar\eta(t=0) &= u, \end{aligned} \tag{E.16}$$

where g is defined by (E.2), and W is a pseudogradient field for J. The right-hand side of (E.16) is locally Lipschitz thanks to (γ). By force of (α) the following estimate analogous to estimate (E.5) is true for all possible $\bar\eta \in X$:

$$\left\| 2\delta g(J(\bar\eta))\frac{W(\bar\eta)}{\|J'(\bar\eta)\|^2} \right\|_* \le 4\delta\|J'(\bar\eta)\|_*^{-1} \le 4\delta C^{-1}.$$

Hence, by Theorem E.2, there is a unique solution $\bar\eta \in X$ to (E.16). We define η analogously to (E.4) and show that it is desired.

Assumption (A) follows from the definition. Assumption (B) is true because the right-hand side of (E.16) is equal to zero identically for $u \notin K_{a,\varepsilon}$ by (E.2).

All we need now is to check (C). As in (E.7), we find using property (β) that

$$\frac{dJ(\eta)}{dt} = \left\langle J'(\eta), \frac{d\eta}{dt} \right\rangle$$
$$= -2\delta\, g(J(\eta)) \left\langle J'(\eta), \frac{W(\eta)}{\|J'(\eta)\|_*^2} \right\rangle \le -2\delta\, g(J(\eta)) \frac{\|J'(\eta)\|_*^2}{\|J'(\eta)\|_*^2} = -2\delta\, g(J(\eta)). \tag{E.17}$$

Analogously to (E.8) and (E.9), relations (E.7) and (E.17) imply

$$J(\eta(1,u)) \le a - \delta$$ □

for all $u \in A_{a+\delta}$.

E.5 The First Mountain Pass Theorem

Let X be a Banach space and $J : X \to \mathbb{R}$ be a functional of the class C^1.

Let θ be the origin of X. Let $\mathcal{P}$ be a family of paths π connecting points θ and $y \in X$, $y \neq \theta$, i.e., a family of continuous functionals $\pi : [0,1] \to X$ with

$$\pi(0) = \theta, \qquad \pi(1) = y.$$

THEOREM E.3. *Let there exist such a neighborhood U of θ with $y \notin U$ that*

$$\max\{J(\theta),\ J(y)\} < \inf_{u \in \partial U} J(u).$$

Then there is a sequence $\{u_n\} \subset X$ such that

$$\begin{aligned} J(u_n) &\to I \qquad \text{in } \mathbb{R}, \\ J'(u_n) &\to 0 \qquad \text{in } X^* \end{aligned} \tag{E.18}$$

as $n \to \infty$, where I is defined by

$$I = \inf_{\pi \in \mathcal{P}} \sup_{t \in [0,1]} J(\pi(t)).$$

Proof. For any path $\pi \in \mathcal{P}$ there exists, because of the compactness of the interval $[0,1]$ and the continuousness of the mapping π, a point u_π such that

$$J(u_\pi) = \max_{u \in \pi} J(u).$$

Because $y \notin U$, any path $\pi \in \mathcal{P}$ intersects ∂U; hence there is an $\varepsilon > 0$ such that, for all $\pi \in \mathcal{P}$,

$$m + 2\varepsilon \le J(u_\pi),$$

where

$$m \equiv \max\{J(\theta),\ J(y)\}.$$

By definition,

$$I = \inf_{\pi \in \mathcal{P}} J(u_\pi),$$

hence, there is a sequence of paths $\{\pi_n\}$ with

$$\lim_{n\to\infty} J(u_{\pi_n}) = I \geq m + 2\varepsilon. \tag{E.19}$$

Suppose, on the contrary, that a sequence satisfying (E.18) does not exist. Then there is a positive constant C such that the inequality

$$\|J'(u)\|_* \geq C \tag{E.20}$$

holds for all $u \in K_{I,\varepsilon}$ with ε from (E.19). This implies that all the assumptions of the Deformation lemma are true and, consequently, there is a mapping η satisfying

$$\eta(0,u) = u \qquad \text{for all } u \in X,$$

$$\eta(t,u) = u \qquad \text{for all } u \notin K_{I,\varepsilon}, \tag{E.21}$$

$$\eta(1, A_{I+\varepsilon/2}) \subset A_{I-\varepsilon/2}. \tag{E.22}$$

In accordance with the inequality in (E.19) there is a path π' such that

$$\max_{u\in\pi'} J(u) \leq I + \varepsilon/2.$$

We may define the mapping $\bar{\pi} : [0,1] \to X$ as

$$\bar{\pi}(t) = \eta(1, \pi(t)). \tag{E.23}$$

The mapping $\bar{\pi}$ is continuous because of the continuousness of π and η. From (E.21) we have

$$\pi(\theta) = 0, \qquad \pi(1) = y,$$

that is, $\bar{\pi} \in \mathcal{P}$.

From (E.22) we have

$$\max_{u\in\bar{\pi}} J(u) \leq I - \varepsilon/2,$$

which contradicts the definition of I. This means that (E.20) can not be true, and hence the theorem is proved. □

E.6 The Second Mountain Pass Theorem

Recall the definition of §13.

DEFINITION. A functional J of the class $C^1(X)$ is called satisfying the PS_c-condition on the Banach space X if, from any sequence $\{u_n\}$ with

$$\begin{aligned} J(u_n) &\to c \qquad \text{in } \mathbb{R}, \\ J'(u_n) &\to 0 \qquad \text{in } X^* \end{aligned}$$

as $n \to \infty$, one can choose a subsequence strongly convergent in X to some of its elements.

We have

THEOREM E.4. *Let the conditions of Theorem E.3 be satisfied. Let a functional J satisfy PS_I on X. Then there is a $u \in X$ such that*

$$\begin{aligned} J(u) &= I, \\ J'(u) &= 0. \end{aligned}$$

Proof. According to Theorem E.3 there is a sequence $\{u_n\} \subset X$ satisfying (E.18):

$$\begin{aligned} J(u_n) &\to I, \\ \|J'(u_n)\|_* &\to 0. \end{aligned}$$

By force of the property PS_I, we may choose a subsequence with

$$u_n \to u \qquad \text{in } X. \tag{E.24}$$

Since functionals J and J' are continuous, (E.18) and (E.24) imply

$$J(u) = I, \qquad \|J'(u)\|_* = 0,$$

which proves the theorem. □

E.7 The Third Mountain Pass Theorem

As was mentioned in proving Theorem E.4, there exists a sequence of paths $\{\pi_n\} \subset \mathcal{P}$ with

$$\lim_{n\to\infty} \max_{u\in\pi_n} J(u) = I. \tag{E.25}$$

For any π_n and $\varepsilon > 0$, define a neighborhood $\pi_{n,\varepsilon}$ in the following way:

$$\pi_{n,\varepsilon} = \{u \in X, \ \|u - \pi_n\| \leq \varepsilon\} \cap K_{I,2\varepsilon},$$

where

$$\|u - \pi_n\| \equiv \inf_{t\in[0,1]} \|u - \pi_n(t)\|.$$

THEOREM E.5. *Let the conditions of Theorem E.4 hold. Then for any sequence of paths* $\{\pi_n\}$ *satisfying (E.25) there exists a sequence* $\{u_n\} \subset X$ *satisfying (E.18) and the relation*

$$\|u_n - \pi_n\| \to 0 \qquad \text{as } n \to \infty.$$

Scheme of Proof. Let, on the contrary, there exist a sequence of paths $\{\pi_n\} \in \mathcal{P}$ satisfying (E.25) and for which the statement of Theorem E.5 is false, that is, for which there are constants $C > 0$ and $\varepsilon > 0$ such that for all n and $u \in \pi_{n,\varepsilon}$,

$$\|J'(u)\|_* > C$$

holds.

In this case we consider, for every such path, the equation

$$\frac{d\bar{\eta}}{dt} = \kappa_n(\bar{\eta}), \qquad \bar{\eta}(t=0) = u,$$

where

$$\kappa_n(u) = \begin{cases} -2\varepsilon\, g(J(u))g_n(u)\frac{W(u)}{\|J'(u)\|_*^2}, & u \in \pi_{n,\varepsilon}, \\ 0, & u \notin \pi_{n,\varepsilon}, \end{cases}$$

g is defined as in (E.2), W is a pseudogradient field defined as in Section E.3, and $g_n : X \to \mathbb{R}$ is some locally Lipschitz functional equaling identically 1 for $u \in \pi_{n,\varepsilon/2}$ and 0 for $u \notin \pi_{n,\varepsilon}$ (one can show that such a functional exists).

Analogously to (E.23) we find $\bar{\pi}_{n_0} \in \mathcal{P}$ of the form

$$\bar{\pi}(t) = \eta(1, \pi_{n_0}(t)),$$

where η is defined by $\bar{\eta}$ in accordance with (E.4), for which the relation

$$\max_{0\le t\le 1} J(\bar{\pi}_{n_0}(t)) < I - \varepsilon/4$$

holds, i.e., we come to the contradiction that proves the theorem. □

References

R. A. Adams

[1] Sobolev Spaces. Acad. Press, New York and San Diego, CA (1975)

K. Ako, T. Kuzano

[1] On bounded solutions of second order elliptic differential equations. J. Fac. Sci. Univ. Tokyo **11** (1964) 29–37

A. Ambrosetti, P. Rabinowitz

[1] Dual variational methods in critical point theory and applications. J. Funct. Anal. **14** (1973) 349–381

V. Benchi, G. Cerami

[1] Positive solutions of some nonlinear elliptic problems in exterior domains. Arch. Rat. Mech. Anal. **99** (1987) 287–300

H. Berestycki, P.-L. Lions

[1] Nonlinear scalar field equations. I. Existence of a ground state. Arch. Rat. Mech. Anal. **82** (1983) 313–346

[2] Nonlinear scalar field equations. II. Existence of infinitely many solutions. Arch. Rat. Mech. Anal. **82** (1983) 347–375

H. Berestycki, P.-L. Lions, L. A. Peletier

[1] On ODE approach to the existence of positive solutions for semilinear problems in $\mathbb{R}^N$. Indiana Univ. Math. Journal **181** (1981) 141–157

M. S. Berger

[1] On the existence and structure of stationary states for a nonlinear Klein-Gordon equation. J. Funct. Anal. **9** (1972) 249–261

[2] Nonlinearity and nonlinear functional analysis. Academic Press, New York and San Diego, CA (1979)

H. J. Brascamp, E. H. Lieb, J. M. Luttinger

[1] A general rearrangement inequality for multiple integrals. J. Funct. Anal. **17** (1974) 227–237

H. Brezis, E. H. Lieb

[1] Minimum action solutions of some vector field equations. Comm. Math. Phys. 1984, **96**, N 1, 97–113

H. Brezis, L. Nirenberg

[1] Positive solutions of nonlinear elliptic equations involving critical Sobolev exponents. Comm. Pure Appl. Math. **36** (1983) 437–477

H. Brezis, L. Veron

[1] Removable singularities of some nonlinear elliptic equations. Arch. Rat. Mech. Anal. **95** (1986) 185–209

C. J. Budd

[1] Application of Shilnikov's theory to semilinear elliptic equations. SIAM J. Math. Anal. **20** (1989) 1065–1080

C. J. Budd, J. Norburg

[1] Semilinear elliptic equations with supercritical growth rate. J. Diff. Eq. **68** (1987) 169–197

M. Cantor

[1] Elliptic operators and the decomposition of tensor fields. Bull. Amer. Math. Soc. **5** (1981) 235–262

S. Chandrasekhar

[1] An introduction to the study of stellar structure. Dover, New York (1957)

D. C. Clark

[1] A variant of the Lyusternik-Shnirel'man theory. Indiana Univ. Math. Jour. **22** (1972) 65–74

S. Coleman, V. Glazer, A. Martin

[1] Action minima among solutions to a class of Euclidean scalar field equations. Comm. Math. Phys. **58** (1978) 211–221

Ding Yi

[1] Non-radial solutions of semilinear elliptic equations. Math. Acta Sci. **10** (1990) 229–239

W. Y. Ding, W. M. Ni

[1] On the existence of positive entire solutions of semilinear elliptic equations in $\mathbb{R}^N$. Arch. Rat. Mech. Anal. **91** (1986) 283–308

N. Dunford, J. T. Schwartz

[1] Linear operators. Vol. 1, Interscience, New York (1958)

R. Engelking

[1] General Topology. PWN, Warsaw (1977)

S. Fucik, A. Kufner

[1] Nonlinear differential equations. Elsevier Sci. Publ. Comp., Amsterdam and London (1980)

N. Fukagai

[1] Existence and uniqueness of entire solutions of second order elliptic equations. Funk. Ekvacioj **29** (1986) 151–165

Y. Furusho

[1] On decaying entire solutions of semilinear elliptic equations. Japan J. Math. **14** (1988) 97–118

Y. Furusho, I. Kuzano

[1] On the existence of positive decaying entire solutions for a class of sublinear elliptic equations. Canadian J. Math. **40** (1988) 1156–1173

B. Gidas, J. Spruck

[1] Global and local behavior of positive solutions of nonlinear elliptic equations. Comm. Pure Appl. Math. **34** (1981) 525–598

B. Gidas, Wei-Ming Ni, L. Nirenberg

[1] Symmetry and related properties via the maximum principle. Comm. Math. Phys. **68** (1979) 209–243

B. Gidas, L. Nirenberg

[1] Symmetry of positive solutions of nonlinear elliptic equations in $\mathbb{R}^N$. Advances in Math. Supplementary Studies **7 A** (1981) 369–402

D. Gilbarg, N. S. Trudinger

[1] Elliptic partial differential equations of second order. Second edition. Springer, Berlin and New York (1983)

C. Jones, T. Kupper

[1] On the infinitely many solutions of a semilinear elliptic equation. SIAM J. Math. Anal. **17** (1986) 803–835

C. Jones, T. Kupper, H. Plaktier

[1] Shooting argument with oscillation for semilinear elliptic radially symmetric equations. Proc. Royal Soc. Edinburgh, Sec A. **108** (1988) 165–180

D. D. Joseph, T. S. Lundgren

[1] Quasilinear Dirichlet problems driven by positive sources. Arch. Rat. Mech. Anal. **49** (1972) 241–269

B. Kawohl

[1] Rearrangements and convexity of level sets in PDE. Springer, Berlin and New York (1985)

J. B. Keller

[1] On solutions of $\Delta u = f(u)$. Comm. Pure Appl. Math. **10** (1957) 503–510

M. A. Krasnoselskii

[1] Topological methods in the theory of nonlinear integral equations. Macmillan, New York and London (1964)

L. D. Kudrjavcev

[1] Embedding theorems for functions defined in unbounded domains. Doklady Akad. Nauk USSR. **153** (1963) 530–535

[2] Embedding theorems for classes of functions defined on the whole space or half-space. I. Matemat. Sbornik **69** (1976) 616–639, II. Matemat. Sbornik **70** (1976) 3–35

I. Kuzin

[1] Existence theorems for nonlinear elliptic problems in $\mathbb{R}^N$ in the class of block-radial functions. Differential Equations (Russian) **30** (1994) 637–646

M. K. Kwong

[1] Uniqueness of positive solution of $\Delta u - u + u^p = 0$ in $\mathbb{R}^N$. Arch. Rat. Mech. Anal. **105** (1989) 243–266

O. A. Ladyzhenskaya, N. N. Ural'ceva
[1] Linear and quasilinear elliptic equations. Academic Press, New York and San Diego, CA (1968)

P.-L. Lions
[1] Symétrie et compacité dans les espaces de Sobolev. J. Funct. Anal. **49** (1982) 315–334
[2] The concentration-compactness principle in the calculus of variations. The locally compact case. Part 1. Ann. Inst. H. Poincaré **1** (1984) 109–145
[3] The concentration-compactness principle in the calculus of variations. The locally compact case. Part 2. Ann. Inst. H. Poincaré **1** (1984) 223–283
[4] On positive solutions of semilinear elliptic equations in unbounded domains. In: Nonlinear diffusion equations and their equilibrium states. Vol. 2. Springer, New York (1988) 85–122
[5] Symmetries and the concentration-compactness method. In: Nonlinear Variational Problems. Pitman, London (1985)

K. McLeod, J. Serrin
[1] Uniqueness of positive solutions of semilinear equations in $\mathbb{R}^N$. Arch. Rat. Mech. Anal. **81** (1983) 981–997

R. C. McOwen
[1] The behaviour of the Laplacian on weighted Sobolev spaces. Comm. Pure Appl. Math. **32** (1979) 783–795

B. Muckenhoupt
[1] Hardy's inequality with weights. Stud. Math. **44** (1972) 31–38

W. M. Ni
[1] On the elliptic equation $\Delta u + K(x)u^{(n+2)/(n-2)} = 0$, its generalizations, and applications in geometry. Indiana Univ. Math. Journal **31** (1982) 493–529

L. Nirenberg, H. F. Walker
[1] The nullspaces of elliptic partial differential equations in $\mathbb{R}^N$. J. Math. Anal. Appl. **42** (1973) 271–301

E. S. Noussair, E. A. Swanson
[1] Positive solutions of semilinear elliptic equations in unbounded domains. J. Diff. Eq. **57** (1985) 349–372
[2] An $L^q(\mathbb{R}^N)$ theory of subcritical semilinear elliptic problems J. Diff. Eq. **84** (1990) 52–61

O. Oleinik
[1] Some asymptotic problems in the theory of partial differential equations, Cambridge Univ. Press, 1996.

B. Opic, A. Kufner
[1] Hardy-type inequalities. Pitman Res. Notes Math. **216** (1990)

R. Ossereman
[1] On the inequality $\Delta u \geq f(u)$. Pacific Jour. Math. **7** (1957) 1641–1647

R. S. Palais
[1] Morse theory on Hilbert manifolds. Topology **2** (1963) 299–340
[2] Lyusternik-Shnirel'man theory on Banach manifolds. Topology **5** (1966) 115–132
[3] Critical point theory and the minimal principle. Proc. Symp. Pure Math. **15** (1970) 185–212

R. S. Palais, S. Smale
[1] A generalized Morse theory. Bull. Amer. Math. Soc. **70** (1964) 165–171

L. A. Peletier, J. Serrin
[1] Uniqueness of a non-negative solution of semilinear equations in $\mathbb{R}^N$. J. Diff. Eq. **61** (1986) 380–397

S. I. Pohozaev
[1] Eigenfunctions of the equation $\Delta u + \lambda f(u) = 0$ nonlinearity. Doklady Akad. Nauk SSSR **165** (1965) 36–39
[2] On eigenfunctions of quasilinear elliptic problems. Matemat. Sbornik **82** (1970) 192–212
[3] On the method of fibering in nonlinear boundary value problems. Proc. Steklov Math. Inst. **192** (1990) 146–163
[4] On elliptic problems in R^N with supercritical exponent of nonlinearity. Matemat. Sbornik **182** (1991) 467–489
[5] On entire solutions of quasilinear elliptic equations. Doklady Akad. Nauk **318** (1991) 815–819
[6] On entire solutions for some class of quasilinear elliptic equations. Doklady Akad. Nauk **318** (1991) 1319–1324
[7] On the asymptotics of entire radial solutions of quasilinear elliptic equations. Doklady Akad. Nauk **320** (1991) 808–813
[8] On entire solutions of semilinear elliptic equations. In: Progress in partial differential equations: elliptic and parabolic problems, Ed. C. Bandle, J. Bemelmans, and M. Chipot. Pitman Research Notes in Math. **266** (1992) 56–59
[9] On entire radial solutions of quasilinear elliptic problems. Proc. Steklov Math. Inst. **204** (1993) 251–273

G. Polya, G. Szego
[1] Isoperimetric inequalities in mathematical physics. Ann. Math. Studies **27** Princeton Univ. Press, Princeton, NJ (1951)

L. S. Pontryagin
[1] Ordinary differential equations. Addison-Wesley, Reading, MA (1962)

P. Rabinowitz
[1] Minimax methods in critical point theory with applications to differential equations. CMBS Regional Conference Series Math. **65** Amer. Math. Soc. Providence, RI (1968) œøœø

R. Reissig, G. Sansone, R. Conti
[1] Qualitative Theorie nichtlinearer Differentialgleichungen. Edizioni Cremonese, Rome, 1963 œøœø

W. Rother
[1] Bifurcation of nonlinear elliptic equations on $\mathbb{R}^N$. Bull. London Math. Soc. **21** (1989) 567–572
G. Sansone
[1] Non-linear differential equations. Pergamon Press, Oxford (1964) œøœø
S. L. Sobolev
[1] Selected questions of the theory of functional spaces and generalized functions. Nauka, Moscow, 1989 œøœø
W. A. Strauss
[1] Existence of solitary waves in higher dimensions. Comm. Math. Phys. **55** (1977) 149–162 œøœø
C. A. Stuart
[1] Bifurcation for Dirichlet problems without eigenvalues. Proc. London Math. Soc. **45** (1982) 169–192 œøœø
H. Triebel
[1] Spaces of Kudrjavcev type. I. Interpolation, embedding and structure. J. Appl. Math. Anal. **56** (1976) 253–277 œøœø
N. S. Trudinger
[1] On Harnack type inequalities and their application to quasilinear elliptic equations. Comm. Pure Appl. Math. **20** (1967) 721–747 œøœø
S. V. Uspenski
[1] On embedding theorems for weighted spaces. Trudy MIAN **61** (1961) 282–303 œøœø
M. M. Vainberg
[1] Variational methods for the study of non-linear operators. Holden-Day, San Francisco, CA (1964)
[2] The variational method and the method of monotonic operators in the theory of non-linear equations. Nauka, Moscow (1972) œøœø
L. Veron
[1] Weak and strong singularities of nonlinear elliptic equations. In: Nonl. Funct. Anal. and Appl. Ed. F. Browder, Proc. Symp. Pure Appl. Math. **45** Amer. Math. Soc. Providence, RI (1986) œøœø
W. Walter
[1] Entire solutions of the differential equation $\Delta u = f(u)$. J. Australia Math. Soc. **30 A** (1981) 366–368 œøœø
Zhu Xi-Ping
[1] Multiple entire solutions of a semilinear elliptic equation. Nonlinear Anal. Theor. Meth. Appl. **12** (1988) 1297–1316 œøœø
Zhu Xi-Ping, Cao Daomin
[1] The concentration compactness principle in nonlinear elliptic equations. Acta Math. Sci. **9** (1989) 307–328 œøœø
K. Yosida
[1] Functional analysis. Springer, Berlin and New York (1965)

Index

9 783764 353230

GPSR Compliance

The European Union's (EU) General Product Safety Regulation (GPSR) is a set of rules that requires consumer products to be safe and our obligations to ensure this.

If you have any concerns about our products, you can contact us on ProductSafety@springernature.com

In case Publisher is established outside the EU, the EU authorized representative is:

Springer Nature Customer Service Center GmbH
Europaplatz 3
69115 Heidelberg, Germany

Batch number: 10577884

Printed by Printforce, the Netherlands